Processing of Complex Sounds by the Auditory System

Proceedings of
a Royal Society Discussion Meeting
held on 4 and 5 December 1991

Organized and edited by
R. P. Carlyon, C. J. Darwin and I. J. Russell

CLARENDON PRESS · OXFORD
1992

Oxford University Press, Walton Street, Oxford OX2 6DP
Oxford   New York   Toronto
Delhi   Bombay   Calcutta   Madras   Karachi
Kuala Lumpur   Singapore   Hong Kong   Tokyo
Nairobi   Dar es Salaam   Cape Town
Melbourne   Auckland
and associated companies in
Berlin   Ibadan   Madrid

Oxford is a trade mark of Oxford University Press

Published in the United States
by Oxford University Press, New York

A catalogue record for this book is available from the British Library

Library of Congress Cataloging in Publication Data
Processing of complex sounds by the auditory system: proceedings of a
Royal Society discussion meeting held on 4 and 5 December, 1991/
organized and edited by R. P. Carylon, C. J. Darwin, and I. J. Russell.
Includes bibliographical references and index.
1. Auditory perception—Congresses.   2. Auditory pathways—Congresses.
3. Auditory cortex—Congresses.   I. Carylon, R. P.
(Robert P.)   II. Darwin, C. J.   III. Russell, I. J.
QP465.P84   1992   152.1'5—dc20   92-28567
ISBN 0-19-852272-X

Printed in Great Britain by
The Alden Press, Oxford

Processing of Complex Sounds
by the Auditory System

Contents

Preface

The demands placed on the auditory system by many everyday situations necessitate very sophisticated forms of processing. Imagine, for example, that you are in a poster session discussing an interesting point of research with the author of a paper—perhaps one of the contributors to this volume. In order to understand what is being said, your auditory system must first break the speaker's voice down into its many frequency components whilst preserving its temporal properties, extract its pitch, and compensate for the reverberation and filtering caused by the room acoustics. To make matters worse, there are a number of other conversations going on near by, each consisting of sounds with very similar frequency spectra and temporal characteristics to your own, and each acoustically mixed with the voice of your increasingly animated conversational partner. The mechanisms involved in all this processing range from active processes in the cochlea, which result in the exquisite frequency selectivity of the auditory system, to the more central processes involved in pitch perception, perceptual sound segregation, and speech perception. It is almost a truism to say that, in order to obtain a complete understanding of the processes involved, one needs to consider data obtained from a wide range of disciplines: accordingly, we have included contributions from the fields of inner-ear biology, auditory physiology, psychophysics, speech perception, and music perception. The papers that make up this volume, then, provide a multi-disciplinary sketch of how the auditory system processes a complex auditory world. Parts of the sketch are missing, but the picture is both fuller and noticeably different from the one that would have been drawn ten years ago.

The papers printed here were first presented at the Royal Society in December 1991. One of the most encouraging aspects of the meeting was the interaction between scientists from different disciplines, and this is reflected in the quality of the individual papers and of the questions that follow. Unusually, for a Royal Society meeting, we took the step of inviting a number of younger scientists to give presentations, the written versions of which are included here. This experiment was an unqualified success, and was greatly helped by a generous contribution towards travelling expenses from B & W Loudspeakers Ltd. We would also like to thank Mary Manning and Christine Johnson of the Royal Society for their help in organizing the meeting, and Simon Gribbin for ensuring the timely publication of this volume. We hope that it will prove to be a useful guide to the progress being made in the many different fields of hearing research, and will be of interest to readers from a wide range of backgrounds.

April 1992

R. P. C.
C. J. D.
I. J. R.

Processing of complex sounds by the auditory system

A Discussion organized and edited by R. P. Carlyon, C. J. Darwin and I. J. Russell

(*Discussion held 4 and 5 December 1991*)

CONTENTS

Phil. Trans. R. Soc. Lond. B (1992) **336**, 293–428
Printed in Great Britain

Auditory processing of complex sounds: an overview

E. F. EVANS

Department of Communication and Neuroscience, Keele University, Keele, Staffordshire ST5 5BG, U.K.

SUMMARY

The past 30 years has seen a remarkable development in our understanding of how the auditory system – particularly the peripheral system – processes complex sounds.

Perhaps the most significant has been our understanding of the mechanisms underlying auditory frequency selectivity and their importance for normal and impaired auditory processing. Physiologically vulnerable cochlear filtering can account for many aspects of our normal and impaired psychophysical frequency selectivity with important consequences for the perception of complex sounds.

For normal hearing, remarkable mechanisms in the organ of Corti, involving enhancement of mechanical tuning (in mammals probably by feedback of electro-mechanically generated energy from the hair cells), produce exquisite tuning, reflected in the tuning properties of cochlear nerve fibres.

Recent comparisons of physiological (cochlear nerve) and psychophysical frequency selectivity in the same species indicate that the ear's overall frequency selectivity can be accounted for by this cochlear filtering, at least in bandwidth terms.

Because this cochlear filtering is physiologically vulnerable, it deteriorates in deleterious conditions of the cochlea – hypoxia, disease, drugs, noise overexposure, mechanical disturbance – and is reflected in impaired psychophysical frequency selectivity. This is a fundamental feature of sensorineural hearing loss of cochlear origin, and is of diagnostic value.

This cochlear filtering, particularly as reflected in the temporal patterns of cochlear fibres to complex sounds, is remarkably robust over a wide range of stimulus levels. Furthermore, cochlear filtering properties are a prime determinant of the 'place' and 'time' coding of frequency at the cochlear nerve level, both of which appear to be involved in pitch perception.

The problem of how the place and time coding of complex sounds is effected over the ear's remarkably wide dynamic range is briefly addressed.

In the auditory brainstem, particularly the dorsal cochlear nucleus, are inhibitory mechanisms responsible for enhancing the spectral and temporal contrasts in complex sounds. These mechanisms are now being dissected neuropharmacologically.

At the cortical level, mechanisms are evident that are capable of abstracting biologically relevant features of complex sounds.

Fundamental studies of how the auditory system encodes and processes complex sounds are vital to promising recent applications in the diagnosis and rehabilitation of the hearing impaired.

1. INTRODUCTION

In the Prado Museum in Madrid are five paintings by Brueghel de Velours, each illustrating one of the senses. The painting on hearing (*El oido*) depicts the problems this volume is concerned with. It shows our ears assailed by competing, complex sounds arising from many different sources: environmental, musical, speech and so on. Our ears have the task of separating the sources and analysing the individual sounds. Both binaural and monaural cues are involved; this conference and hence this overview concentrates on monaural processing.

If we take speech (figure 1) as a paradigm of a complex sound, it is characterized by bands of energy spanning multiple frequencies (e.g. the consonants and vowel of figure 1) and changing in frequency and amplitude with time (particularly the component

frequencies – formants – of the vowel). The ear has at least three tasks to perform on a complex sound. First, to separate out the individual frequency components from several, simultaneously present (for example the vowel formants). This is termed frequency selectivity (or frequency analysis or resolution): the ability of the ear to resolve (breakdown) a complex sound into its individual, component frequencies. The spectral complexity of the stimulus, thus analysed, is the main determinant of its perceived timbre. It is also important, together with temporal cues, for determining the pitch of a complex sound (for speech, the perceived laryngeal frequency) and this information is essential in enabling the auditory system to differentiate between speakers. The second task for the ear is to enhance the spectral and temporal contrasts of the resolved frequency components in order to compensate for poor signal-to-noise-ratios in naturally occur-

Phil. Trans. R. Soc. Lond. B (1992) **336**, 295–306
Printed in Great Britain

295

[1]

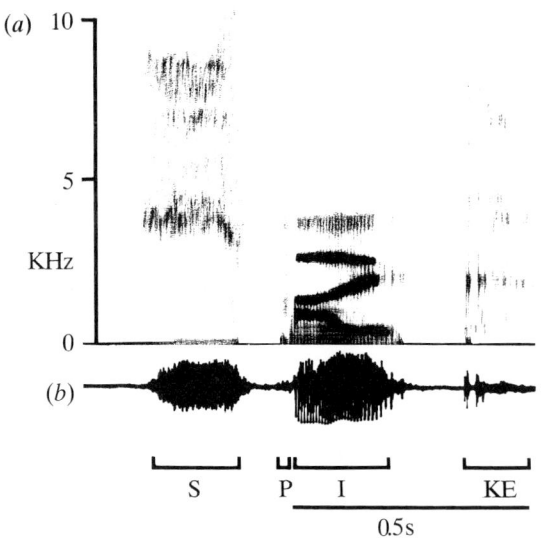

Figure 1. Frequency and time representation of the word 'spike'. (*a*) Spectrogram; (*b*) waveform. (From Evans 1974*a*.)

highly schematic summary diagram of the major components of the monaural auditory system.

1. The outer ear and middle ear act as signal 'conditioners', i.e. they emphasize those frequencies that are of most relevance for the particular species: for example, for humans about 1–3 kHz, for cats and guinea-pigs, about 5–15 kHz.

2. The cochlear partition in the inner ear or cochlea. This is the frequency analyser of the auditory system. The basilar membrane and the organ of Corti together perform the function of a distributed filter bank from which tuned channels emerge in the form of the responses of the cochlear nerve fibres.

3. The cochlear nerve fibres encode the filtered responses in terms of their spatio-temporal patterns of discharges in a cochleotopic ('tonotopic') map of activity across the array of tuned fibres emanating from the apical (low-frequency) to basal (high-frequency) end of the cochlea.

4. These filtered responses are passed on, in parallel, to a number of subsystems within the brainstem nuclei: starting with the cochlear nucleus (CN) where between two (shown here) or more likely three or more independent processing pathways diverge.

5. How far the subsystems are kept separate as we ascend the auditory midbrain is not clear. Nor do we have as clear a picture of the segregation function in the auditory cortex as in the visual system (except perhaps in the bat, thanks to Suga (this symposium)).

Unlike Brueghel, I have time only to paint a picture of the auditory system using rather broad brush

ring sounds, and to aid the third taks: namely to extract and abstract the behaviourally meaningful cues from the results of the peripheral spectral analysis. As far as behaviourally meaningful complex sounds like speech are concerned, this includes determining the spacing of the frequency components and their changes in time.

Where is all this carried out? Figure 2 shows a

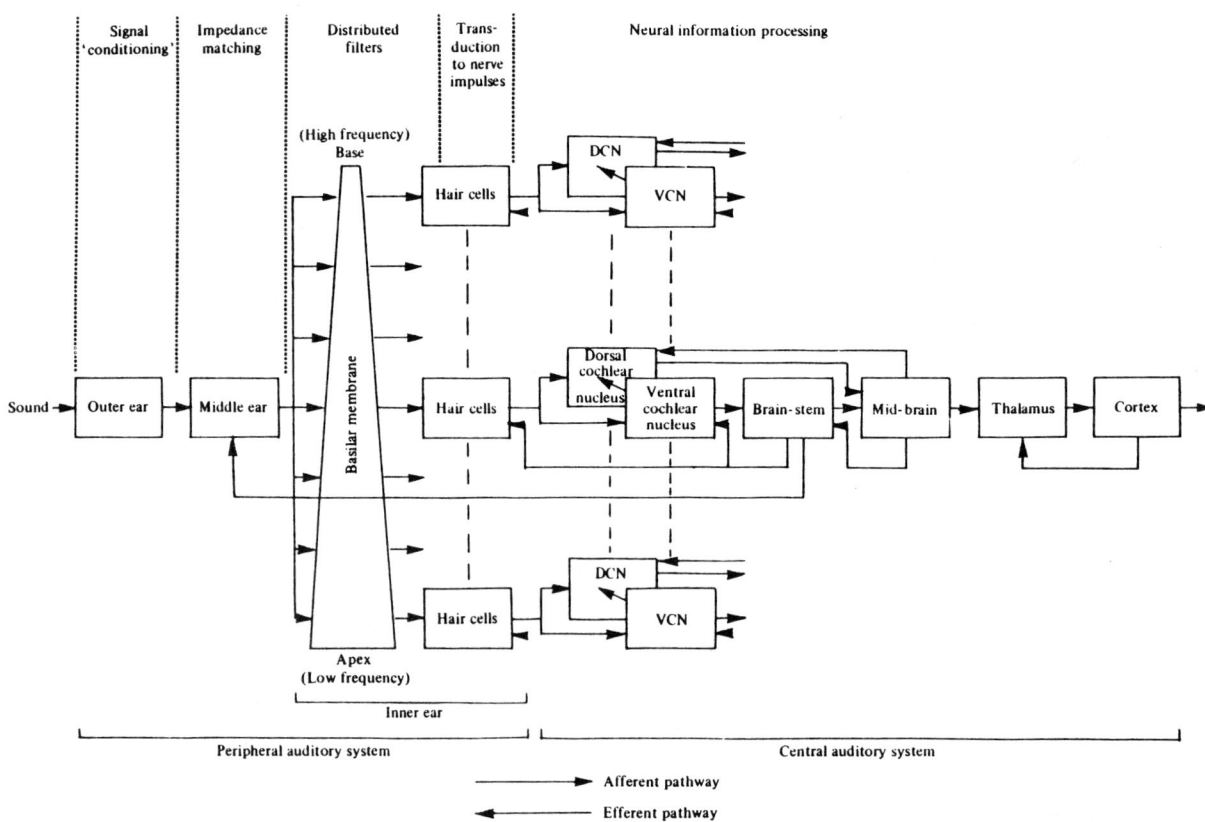

Figure 2. Schematic anatomical and functional map of the auditory sytem. (From Evans 1982.)

strokes. I will restrict my painting to a rather personal caricature of the three most clearly identified functions of the monaural auditory system at the three sites that have interested me over the last 30 years: the cochlear nerve, cochlear nucleus and auditory cortex. In so doing, I hope to emphasize a striking contrast in response properties that exists between cochlear nerve and cortical levels. To use a musical analogy, the response characteristics of the auditory periphery resemble those of a piano; of the cortex, those of a symphony orchestra!

2. PERIPHERAL FREQUENCY ANALYSIS SEEN AT THE LEVEL OF THE COCHLEAR NERVE

Central to our understanding of how the ear analyses complex sounds at this level into their component frequencies (frequency selectivity), is the physiological tuning curve of an individual fibre in the cochlear nerve (continuous curve in figure 3). The frequency threshold (or tuning) curve (FTC) describes the threshold sound intensity required to elicit a minimal response from the cochlear nerve fibre (an increase in the spike discharge rate) as a function of frequency. Under normal conditions (curve A in figure 3), the response of cochlear nerve fibres is very sharply tuned: they represent exquisitely tuned filters with bandwidths of the order of one sixth of an octave and cut-off slopes of several hundred dB per octave (see Evans (1975a) for review).

In the early 1970s I showed that this tuning was physiologically vulnerable. For example, a reduction in the oxygen supply to the cochlea could change the

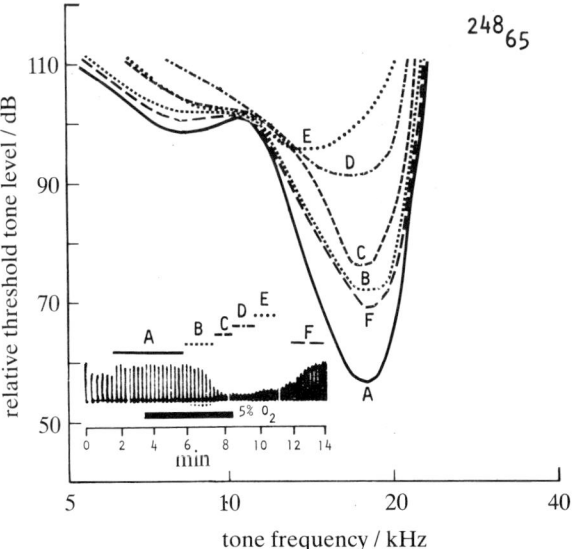

Figure 3. The effect of hypoxia on the tuning of a single cochlear nerve fibre. Curve A: frequency threshold curve (FTC) of cat cochlear nerve fibre. Curves B, C, D, E and F: FTCs obtained during and following 4 min hypoxia produced by reducing the expired air oxygen concentration to 5%. Inset shows timing of collection of FTC data in relation to the period of hypoxia and the brief reversible reduction in amplitude of the gross cochlear action potential in response to a fixed amplitude click. (From Evans 1974*b*.)

tuning from sharp and low threshold (curve A in figure 3), progressively (curves B–D) to blunt and high threshold (curve E). This was one of the lines of evidence that led to the proposal (Evans 1972) of the existence of an additional, biologically active (i.e. physiologically vulnerable) process sharpening up the relatively poorly tuned (passive) mechanics of the basilar membrane. We called it the cochlear 'second filter' as an expression of ignorance of the underlying mechanisms, but favoured (Evans 1972) a positive feedback process first proposed by Gold (1948; admittedly on spurious psychophysical evidence: see Evans (1975a)). At the time, all the evidence (poor basilar membrane tuning) pointed against a 'parallel' filtering enhancement process but supported a 'serial' process (see Evans & Wilson, 1973, 1975). The nature of the 'second filter' hypothesis has been much misunderstood and it perhaps needs to be emphasised that in its original framing it was sufficiently wide to embrace both 'parallel' and 'serial' filtering processes. Indeed, the validity of the extant mechanical data was specifically questioned (Evans 1972, 1975a). Furthermore, our studies of the influence of kanamycin poisoning of the cochlea (Evans & Harrison 1976; Harrison & Evans 1977) provided clear evidence that the integrity of the outer hair cells was crucial to sharp tuning of the cochlear nerve fibres emanating from the inner hair cells. For the mammalian cochlea, it now looks as though a parallel, physiologically active, enhancement process is required, involving the outer hair cells as the producers of motile energy coupling back to the inner hair cells (and to the basilar membrane, thus imparting sharp tuning to its mechanics: Sellick *et al.* (1982) in the guinea pig; Khanna & Leonard (1982) in the cat). On the other hand, for reptilian inner ears (where basilar membrane tuning is poor or negligible), a serial process appears to be involved, with electromechanical tuning of the hair cells themselves: see, for example, Crawford & Fettiplace (1981).

These frequency threshold curves can be considered to represent filters passing most energy at the characteristic frequency (CF): the most sensitive frequency, and least energy at frequencies further away. One of the remarkable things about these filters is that in spite of their resulting from nonlinear processes in the cochlea, they act surprisingly linearly in response to complex, i.e. multicomponent stimuli. In other words, their responses to these stimuli can be predicted to a first approximation over a surprisingly large dynamic range, by models containing linear band-pass filters having the same amplitude characteristics as the FTC (see Evans (1975a, 1989a) for reviews; see also Evans (1977, 1981, 1985a)).

Wilson and I first showed 20 years ago (Wilson & Evans 1971; Evans & Wilson 1973) that the characteristics of these physiological filters in the cat accounted approximately, but not exactly, for several aspects of psychophysical human frequency selectivity. Since then, Pickles (e.g. 1975) has questioned this conclusion, based on comparisons between his behavioural critical band data and our cochlear fibre data, both in the cat. Recently, I have had the opportunity

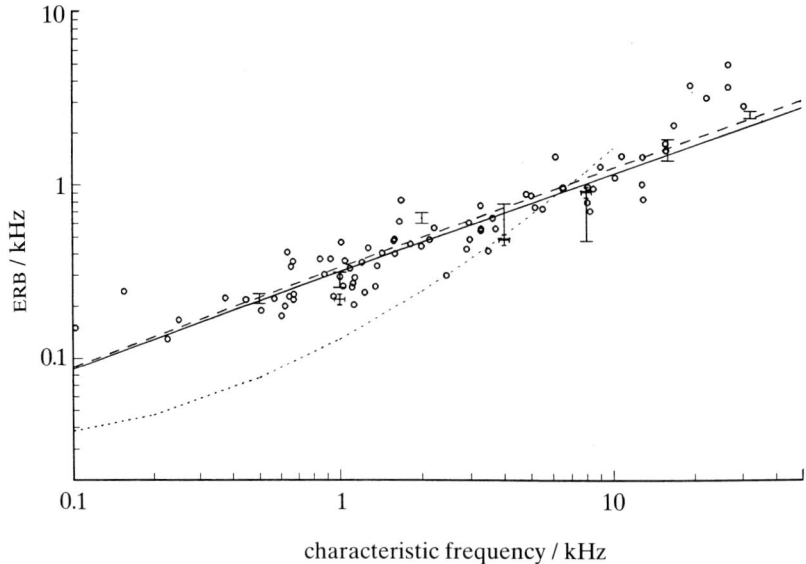

Figure 4. Comparison of behavioural and physiological frequency selectivity in the guinea-pig. Open circles: equivalent rectangular bandwidths (ERBs) of guinea-pig cochlear nerve fibres recorded under optimal conditions, i.e. with systemic arterial blood pressure 60 mm hg and gross cochlear action potential threshold 35 dB p.e. SPL. Bracket symbols: behavioural equivalent rectangular bandwidths derived from comb-filtered noise masking ± 1 s.e. Star points: behavioural equivalent rectangular bandwidths derived from brandstop noise masking. The continuous line is the regression line through the mean equivalent rectangular bandwidths determined by comb-filtered noise masking. The dashed line is the regression line through the physiological data points. The dotted line represents human equivalent rectangular bandwidths from a variety of sources summarised in Moore *et al.* (1983). (From Evans *et al.* 1989, 1992.)

to make more appropriate comparisons between physiological and psychophysical selectivity in the guinea-pig (Evans et al. 1989, 1992). To make these comparisons, we make the assumption (mentioned above) that the physiological FTCs represent linear filters, and express the bandwidth of these filters as the width of a rectangular filter having the same area when the FTCs are plotted on linear power, linear frequency coordinates: the equivalent rectangular bandwidth (ERB). Figure 4 shows the excellent agreement obtained between the ERBs of 80 single fibres in the guinea-pig cochlear nerve (open circles) compared with behavioural measurements of frequency selectivity in the same species, using comb-filtered noise masking and band-stop (notch) noise masking (bracket and cross symbols respectively). The good correspondence between the physiological and psychophysical data led us to hypothesize that the ear's overall frequency selectivity was already largely determined by peripheral mechanisms: by the exquisite tuning of the cochlea. An important consequence of this, and of the physiological vulnerability of cochlear tuning, is that when the function of the cochlea is impaired, through disease, drugs, mechanical or surgical damage, the ear's overall (psychophysical) frequency selectivity is likewise impaired (see Evans (1975b, 1978b) for review), and tests of frequency selectivity may, under certain conditions, be more sensitive tests of damage to the inner ear than conventional tests (e.g. West & Evans 1990).

Thus, in principle at least, the distributed filter bank of cochlear nerve fibres maps peaks of energy in the spectrum of the incoming sound into peaks of activity in the array of cochlear nerve fibres emanating from the cochlea. This is the place representation of a complex sound. However, as has been indicated, in impairment of cochlear function, deterioration in cochlear tuning will produce a blurred place representation of the spectrum of the incoming sound. This is why linear amplification, as in conventional hearing aids, though making the speech audible, will not make it any clearer (Evans 1978b; Pick & Evans 1982).

Similar blurring of the place representation occurs as a result of a severe nonlinearity in normal cochlear nerve fibre responses, namely the restricted dynamic range of most fibres. The range of sound intensities capable of eliciting changes in discharge rate with change in sound level is limited to about 40 dB in most cochlear fibres (Sachs & Abbas 1974; quantified as to proportions in Evans & Palmer (1980)). Because of this restriction, the mapping of activity in terms of discharge rate across place in the responses of the majority of cochlear nerve fibres is blurred out completely at moderate to high sound pressure levels (see Evans (1981) for review). However, Sachs & Young (1979), and Palmer & Evans (1979) have shown, in the cat, and Winter *et al.* (1990) in the guinea-pig, that this is not true of all cochlear nerve fibres: a small minority have (somewhat) higher thresholds than average and much wider dynamic ranges (not less than 70 dB), so that they are in principle capable of representing the spectra of complex stimuli on a rate place basis. This subset of cochlear nerve fibres account for some 5% of the total population: they are a sub-population of fibres having the lowest or no spontaneous discharge rates (Evans &

Palmer 1980; Schalk & Sachs 1980). Liberman (1978) was the first to show that the cochlear nerve is not a homogeneus population of fibres: at least two (or three) sub-populations exist: a small low spontaneous rate population having higher thresholds and wider dynamic ranges; the remainder having the higher spontaneous rates, lower thresholds and dynamic ranges restricted to 40 dB on average (Liberman 1978; Evans & Palmer 1980; Schalk & Sachs 1980).

This minority population is capable of acting as rate place encoders of absolute spectral level at moderate to high stimulus levels. Although this situation can be adequate for models of intensity coding (Viemeister 1988), it does not seem very parsimonious for the great majority of cochlear nerve fibres not to be involved in coding the spectra of complex sounds at moderate to high levels. On the other hand, the presence of background noise can shift the dynamic ranges of cochlear nerve fibres (Evans 1974*b*; Costa-lupes *et al.* 1984), particularly when the effects of the descending, efferent projections to the cochlea are taken into account in the unanaesthetized preparation (May & Sachs 1992).

Looking for alternatives to rate processing, the potential for encoding complex sounds over a wide dynamic range has been explored in terms of the temporal discharge patterns of cochlear fibres. The important point here is that the timing of cochlear nerve fibre discharges to frequencies up to about

5 kHz or so is related to the period or multiples of the period of the stimulus. This is termed the 'phase locking' of the discharges to the stimulus waveform, and represents the basis of time representation theories of encoding frequency.

By using measures of the degree of phase locking as a function of the frequencies corresponding to the place of activity, Young & Sachs (1979) showed very beautifully that it was possible to extract the spectrum of speech vowels from the temporal discharge patterns of cochlear nerve fibres over a very wide dynamic range, including sound pressure levels well above those at which the discharge rate profiles of most were saturated and blurred out. Whether or not the higher levels of the auditory system can extract this representation, however, is still not clear.

What is clear however, is that the tuning of cochlear nerve fibres is surprisingly well represented in the temporal discharge patterns of cochlear nerve fibres over a very wide dynamic range (figure 5). These are the plots of the tuning of three cochlear nerve fibres (having relatively low, medium and high CFs) at low (lower half of figure 5) and high (upper half) intensities, obtained with a variety of complex stimuli. The measures of tuning have been extracted from the weighting or degree of phase-locking of the cochlear nerve fibre to individual component frequencies of a complex stimulus generated either by click trains (plus symbols), tone complexes (open circles) or broadband

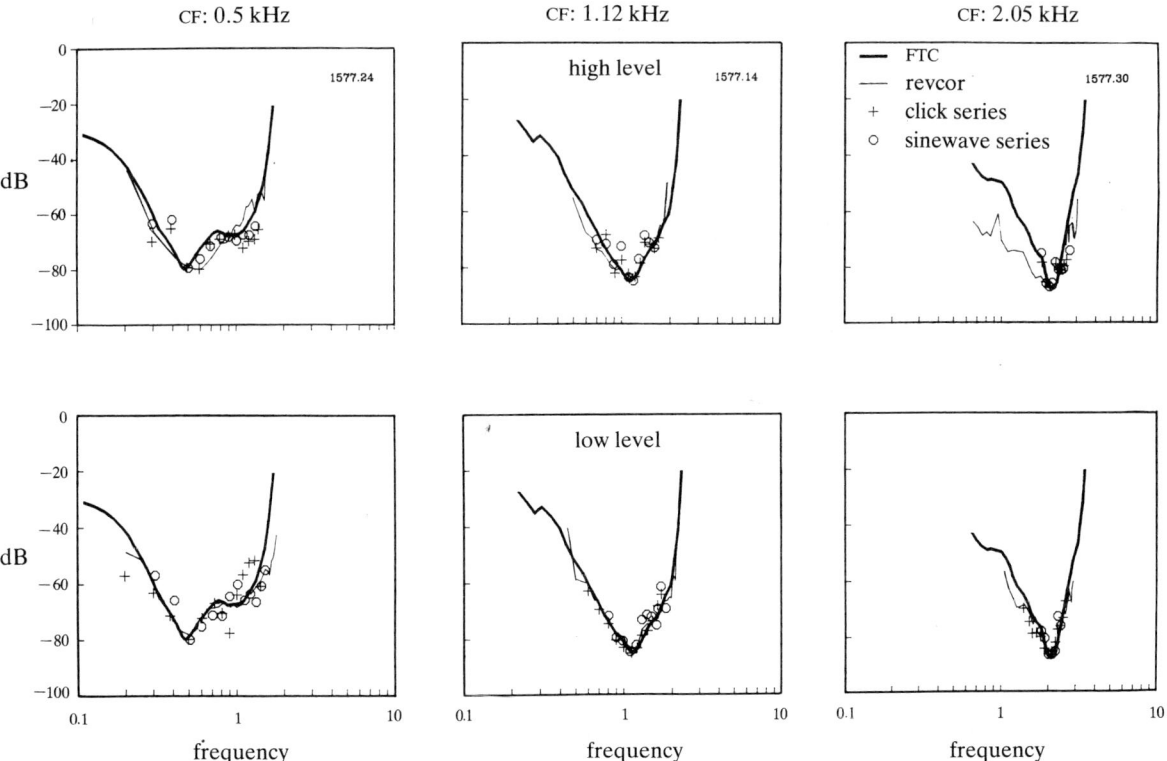

Figure 5. Derivations of cochlear filtering through the discharge patterns of cat cochlear nerve fibres of low, middle and higher characteristic frequency near threshold (lower half) and at stimulus levels 50 dB or more above threshold (upper half). Thick lines: pure tone frequency threshold curves; thin lines: filter functions derived from reverse correlograms with broad-band noise stimulation; crosses and circles: filter functions derived from weighting measures from temporal synchrony in response to click trains and sine-phase mixed tone complexes, respectively. (From Evans 1989*a*.)

noise (thin line). (In the case of broadband noise, the filtering was extracted by the reverse correlation technique of De Boer (1969); see Evans 1977, 1985*a*). Superimposed is the pure tone FTC (thick line). The filtering remains amazingly well preserved in the weighting of the time pattern over a surprisingly large dynamic range: as great as 90 dB in the cat (Evans 1977, 1981). (In the guinea-pig, however, there is evidence from our own and other studies that there are greater nonlinearities with level (Harrison & Evans 1982; Cooper & Evans 1988), as appears also to be the case in the rat (Moller 1977).)

One question which has been of great interest is how important these temporal patterns of spike discharge are for the determination of pitch. There has been much debate over the years as to which of the 'place' or 'time' neural representations is most relevant. My own studies of attempts by others to disprove the importance of time cues in pitch perception have led on the one hand to a realisation of the importance of cochlear tuning in the determination of the temporal patterns of spike discharges and on the other, to the position that it seems inescapable that both place and time cues must be involved in the perception of pitch independently or together, each assuming different importance depending upon the stimulus. Figure 6*a* shows stimuli (originally devised by Patterson (1973)) but used by Wightman & Green (1974) to test the importance of time cues in pitch perception. The two stimuli in figure 6*a* are the waveforms generated by summing the fifth to the tenth harmonics of a 200 Hz fundamental mixed in cosine phase (left) and random phase (right). The former waveform is very periodic, the other aperiodic. Wightman & Green argued that the cosine-mixed waveform could be expected to produce highly synchronized discharges in cochlear fibres by virtue of the

waveform periodicity; the other not so. And yet the pitches and pitch strengths evoked by the two stimuli are virtually identical. Wightman & Green therefore argued that time cues were not likely to be involved in the perception of the pitch. However, the argument entirely ignored the role of cochlear filtering. If the two waveforms are passed through a cochlear-like filter, one gets the waveforms shown in figure 6*b*. Not surprisingly, the temporal discharge patterns of cochlear nerve fibres evoked by the very different stimulus waveforms of the upper row are virtually identical. In figure 6*c* are shown inter-spike interval histograms obtained from a single cochlear nerve fibre having a CF at the centre frequency of the harmonic complex under stimulation with the cosine- and random-phase mixed harmonics, respectively. This also demonstrates how even temporal representations of aspects of complex sounds are determined by the peripheral filtering mechanisms, and are therefore importantly correlated with the place of origin.

I have not found a pitch-evoking stimulus where the pitch heard could not be predicted from the fine-time structure of the cochlear nerve fibre discharges as seen in the inter-spike interval histogram or autocorrelogram analyses of the discharge patterns of cochlear fibres in the dominant region of CFs (Evans 1978*a*, 1983, 1986, 1989*b,c*). I have long gone on record that the brain is likely to use both place and time cues for pitch: either will suffice; but when both are present and congruent, then the most salient and discriminable pitch will be heard (Evans 1978*a*; 1989*c*).

3. SPECTRAL AND TEMPORAL CONTRAST ENHANCEMENT: THE COCHLEAR NUCLEUS LEVEL

It is at this level that the cochlear nerve input diverges

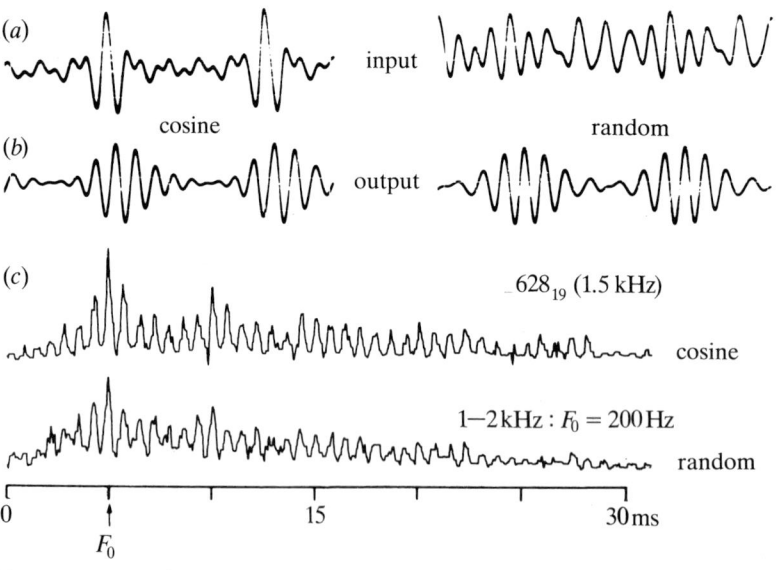

Figure 6. Phase effects are minimised by cochlear filtering. The waveforms in (*a*) represent the result of summing the fifth to the tenth harmonics of a common fundamental in cosine (left) and random phase (right) respectively. (*b*) Effect of filtering the above signals by a band-pass filter with half power bandwidth approximately equal to neural bandwidths of appropriate frequency. (*c*) Inter-spike interval histograms of a single cochlear nerve fibre in response to the signals in (*a*), in which the harmonics are evenly distributed across the cochlear fibre's filter function (CF: 1.5 kHz); fundamental frequency of complex: 200 Hz. (From Evans 1978*a*.)

[6]

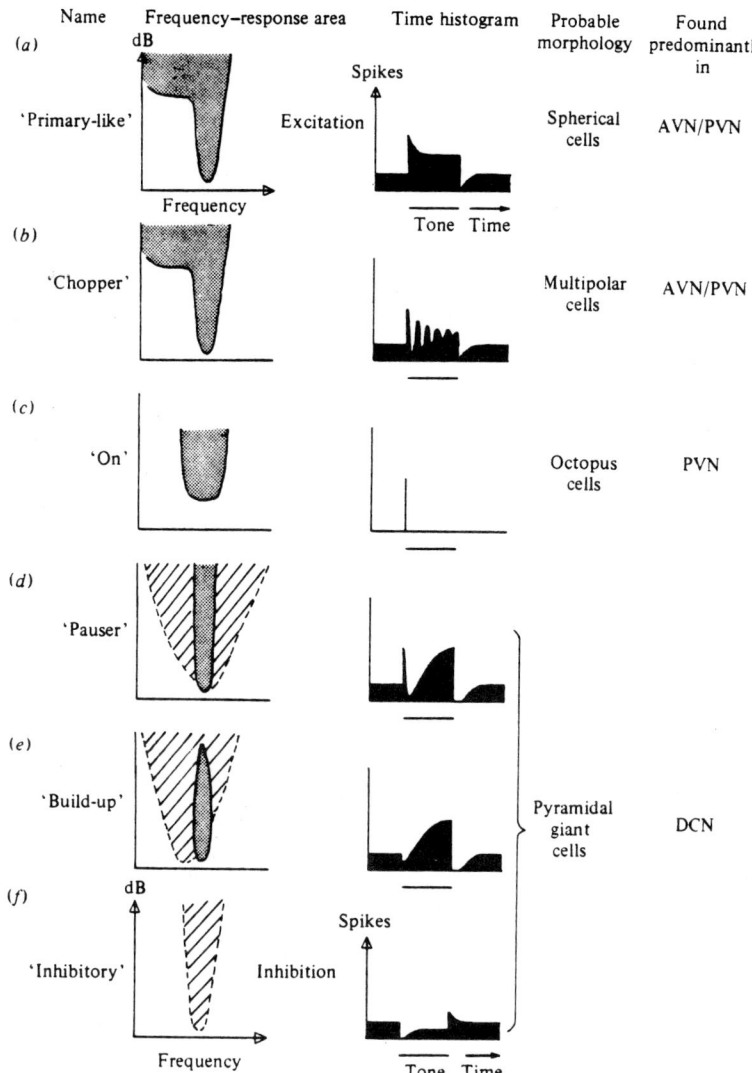

Figure 7. Diagram of frequency and time responses of major types of cells in cochlear nucleus. Left-hand column: diagrams of frequency response areas (axes as in (*f*)) indicating typical excitatory (stippled) and inhibitory (shaded) frequency-response areas of each type of cell. Inhibition alone (*f*) or inhibition combined with excitation (*d, e*) is found predominantly in the morphologically more complex dorsal cochlear nucleus (DCN). Middle column: schematic diagrams of peristimulus-time histograms of response to characteristic frequency tone of each type of cell (axes as in (*f*). The shape of the PST histogram gives rise to the 'nick-name' of each cell type to the left of the left hand column. Right-hand column: probable cell types and location. (From Evans 1982.)

to innervate a wide variety of types of cell (see diagrammatical and simplified representation in figure 7) in terms of location, morphology, receptive field organisation and response as a function of time before, during and following stimulation (see Evans (1975*a*) and Young (1985) for reviews).

A great deal of interest has centred on the class of cells found in the dorsal division of the cochlear nucleus (DCN) that exhibit extensive lateral inhibition of their response at frequencies above and below a narrow or barely present excitatory area: the type IV cells of Evans & Nelson (1973), and Young and colleagues (e.g. Young 1985), and figure 7*e*.

What is the function of this lateral inhibition? In recent experiments (Zhao & Evans 1990; Evans & Zhao 1992*a*), we have been able to demonstrate that this lateral inhibition is glycinergic: comparing the responses before and after blockage by the glycine

antagonist, strychnine, allows us to dissect out the contribution made by the inhibition (figure 8). In figure 8*a* are the automatically mapped receptive fields of a type IV cell in the guinea-pig cochlear nucleus. The two left-hand maps indicate in the height of the small bars, the spike count in response to a tone burst of different (randomized) frequencies and levels indicated by the centre of each bar. The left-hand map is the control, and shows, in the 'white' patches, the lateral inhibition, surrounding barely discernable patches of excitation (between 11 and 12 kHz). Blocking the inhibition with strychnine (middle column) reveals the true extent of the excitatory input to the cell. By subtraction of these receptive fields, we can also derive the extent of the inhibitory field, as shown in the 'three-dimensional' contour plot on the right. It extends throughout the excitatory receptive field, and in fact is strongest at the excitatory

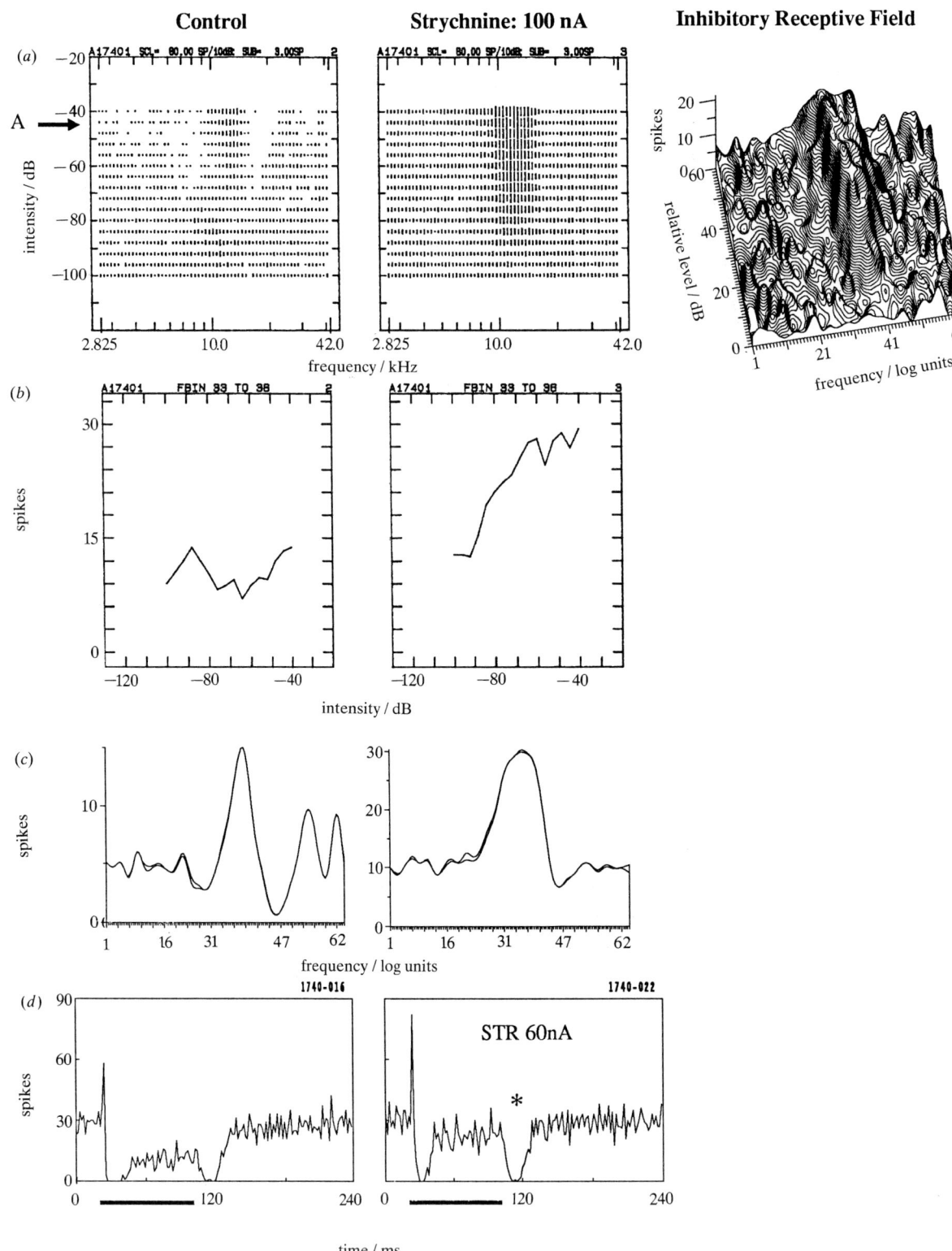

Figure 8. Role of 'lateral' inhibition in determining the response properties of dorsal cochlear nucleus (type IV) cells. Left-hand column shows control analyses; middle and right-hand columns show identical analyses after blocking the inhibition by the iontophoretic application of 100 nA strychnine. (*a*) Receptive field maps; (right) three-dimensional contour map of the strychnine-blocked inhibition, obtained by subtracting the control receptive field map from that obtained under strychnine. (*b*) Rate-level functions obtained at CF (12 kHz), from the receptive field maps in (*a*). Note conversion by strychnine blockade of non-monotonic rate-level function (control) into a steep monotonic function. (*c*) Iso-level functions taken at 50 dB above threshold (−44 dB) indicated by the arrow A in (*a*). Note extensive inhibitory side-bands in control case, surrounding narrowed excitatory response region, and removed by the strychnine blockade (right) with widening of the excitatory area and reduction in response contrasts. (*d*) Peristimulus-time histograms of the cell at 8 kHz, i.e. in the inhibitory side-band below CF. Note strychnine blockade of the sustained inhibition only, leaving unaffected the transient inhibition at the on-set of the stimulus, and off-inhibition (indicated by asterisk). Bar indicates tone. (From Evans & Zhao 1992*c*.)

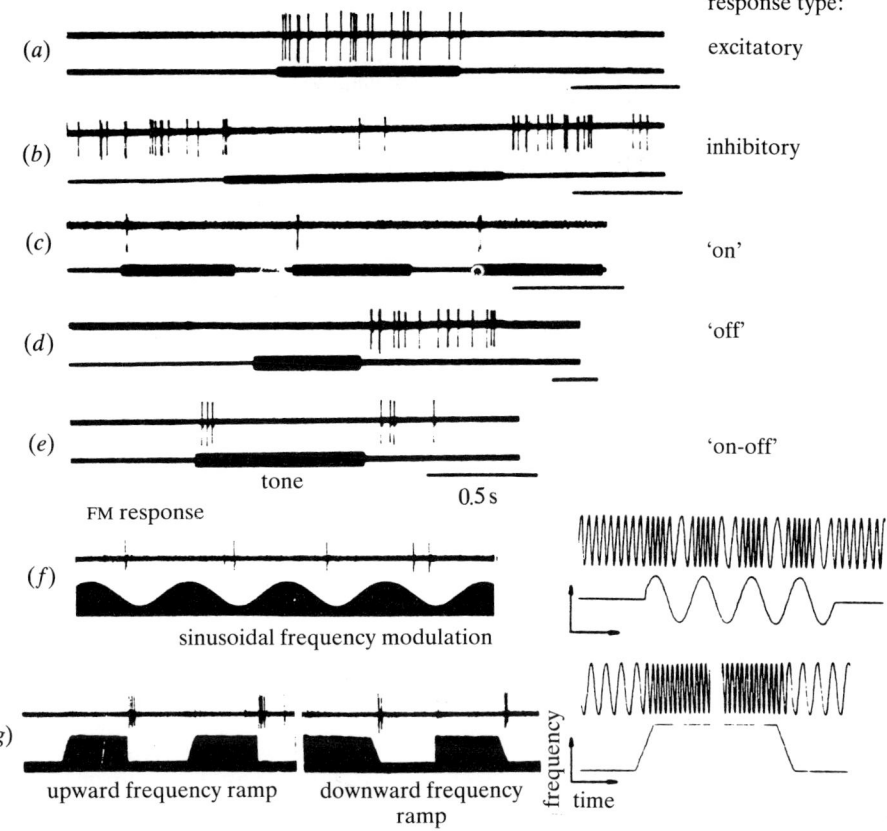

Figure 9. Variety of response types of cells in primary auditory cortex. (*a–e*) Types of response to steady tones. (*f, g*) Cell responding to frequency-modulated, not steady tones. Note response selectivity to direction of frequency sweep: in the downward direction, not to upward sweeps. Black envelope indicates excursions of frequencies illustrated by the waveforms to the right. (From Evans 1968, 1982.)

CF. (Thus, the term *lateral* inhibition is misleading.) Figure 8*b* shows the rate-level functions at the CF. Strychnine blockade of the inhibition converts the non-monotonic rate-level function characteristic of type IV cells into a monotonic function (right-hand plot) characteristic of primary afferents. Figure 8*c* shows the corresponding iso-level functions, i.e. response as a function of frequency at the constant level indicated by the arrow at A in figure 8*a*. In the control case (left plot), inhibitory 'troughs' surround the narrow excitatory response centre. Strychnine blockade of the inhibition (right-hand plot) eliminates the inhibitory troughs and reveals the true width of the excitatory receptive field. Thus, we can see that the 'lateral' inhibition is responsible for narrowing the excitatory response area and enhancing spectral contrasts. Young (this symposium) will show extreme examples of this function in connection with his hypothesis that the DCN is concerned with the analysis of the sharp spectral notches generated by reflections within the pinna. As we have shown elsewhere (Palmer & Evans 1982), 'lateral' inhibition can also bias a cell's 'working-point' to extend and optimize its dynamic range.

But not only spectral contrasts but temporal contrasts are further enhanced at the cochlear nucleus level. Off-inhibition (the suppression of discharge immediately following the offset of a stimulus, whether excitatory or, as in figure 8*d*, inhibitory) serves to enhance the response contrast between the stimulus

'on' and the stimulus 'off'. (As yet, we have not been able to determine what transmitter, if any, is responsible: strychnine blockade leaves the off-inhibition unaffected as shown at the asterix.) This enhancement of temporal contrast improves the modulation of discharge of cochlear nucleus cells in response to amplitude modulated stimuli, compared with cochlear nerve fibres, as was first shown by Moller (1972).

We have recently shown that there are several different types of inhibition in the cochlear nucleus, responsible for different aspects of spectral and temporal contrast enhancement, each apparently under control of different neurotransmitter systems and therefore capable in principle of exquisitely fine control by other parts of the auditory system (Evans & Zhao 1992*b,c*).

4. ABSTRACTION OF SALIENT FEATURES OF COMPLEX STIMULI: AUDITORY CORTEX

Compared with the cochlear nerve and even the cochlear nucleus where responses are relatively straight-forward representations of stimulus spectral content, the responses obtained at the level of the auditory cortex are a very mixed bag (figure 9). 'On' responses, inhibitory responses, 'off', 'on–off' responses are very common to pure tone stimuli. Even in the primary auditory cortex of the unanaesthetised preparation the cells are difficult to drive reliably by pure tones. Many cells respond much more vigorously and consistently to complex sounds. In the case of the cat, the most effective stimuli (used as typical 'search stimuli') are what I have termed 'backdoor noises': the sounds that are heard outside the average British backdoor around midnight calling the pet feline in for the night: kissing, hissing noises, jangling of keys, etc. (Evans 1968)!

Of particular interest for complex signal analysis, are the small minority of cells that do not respond to pure tones at all, but do respond selectively to complex patternings of frequency and amplitude in time (figure 9*f,g*). Thus, there are cells specifically selective to frequency changes which are capable of signalling both the direction (e.g. to downward frequency transitions in the single cell illustrated in figure 9*f,g*) and the rate of frequency change. These were first described by Bogdanski & Galambos (1960) but independently and in more detail by Evans & Whitfield (1964), Whitfield & Evans (1965); and by Suga (1964).

Cells in the uppermost levels of the auditory system are therefore capable of extracting or abstracting salient features of a complex stimulus: is it on, is it off, has it just turned on or off? Is the frequency changing, if so in what direction, at what rate? And so on.

These findings have been extended beautifully in the discovery of cortical cells selective in their response to features of species-specific vocalizations: by the late Peter Winter and colleagues (e.g. Winter & Funkenstein 1971) in the squirrel monkey, and by Suga (this symposium) in the bat.

5. EPILOGUE

None of these considerations that this volume addresses are of purely academic interest. Some of us have interests in developing neural prostheses for restoring some form of hearing to the profoundly hearing impaired, for example with multi-channel cochlear implants (see Evans (1985*b*) for review). We need detailed knowledge on the way in which complex sounds are encoded in the auditory periphery, so that we may mimic these patterns as accurately as feasible, by artificial, electrical, stimulation of the auditory periphery either at cochlear or brainstem levels. Currently, this artificial stimulation is subject to severe technological limitations, particularly at the electrode-biological interface. It is equally important therefore to know how the central levels of the auditory system might deal with this imperfectly as well as with normally encoded information.

REFERENCES

de Boer, E. 1969 Reverse correlation II. Initiation of nerve impulses in the inner ear. *Proc. Konikl. Nederl. Akademie van Wetenshappen – series C Amsterdam* **722**, 129–151.

Bogdanski, D.F. & Galambos, R. 1960 Studies of the auditory system with implanted electrodes. In *Neural mechanisms of auditory and vestibular systems* (ed. G. L. Rasmussen & W. F. Windle), pp. 143–148. Springfield: Charles C. Thomas.

Crawford, A.C. & Fettiplace, R. 1981 An electrical tuning mechanism in turtle cochlear hair cells. *J. Physiol., Lond.* **212**, 377–412.

Costalupes, J.A., Young, E.D. & Gibson, D.J. 1984 Effects of continuous noise on rate response of auditory nerve fibers in the cat. *J. Neurophysiol.* **51**, 1326–1344.

Cooper, N.P. & Evans, E.F. 1988 Comparisons between frequency selectivity manifested in temporal discharge patterns and frequency threshold curves of cochlear nerve fibres in guinea-pig. *Br. J. Audiol.* **22**, 135–136.

Evans, E.F. 1968 Cortical representation. In *Hearing mechanisms in vertebrates* (ed. A. V. S. de Reuck & J. Knight), pp. 272–287. London: J. & A. Churchill.

Evans, E.F. 1972 The frequency response and other properties of single fibres in the guinea pig cochlear nerve. *J. Physiol., Lond.* **226**, 263–287.

Evans, E.F. 1974*a* Neural processes for the detection of acoustic patterns and for sound localization. In *The neurosciences: third study program* (ed. F. O. Schmitt & F. G. Worden), pp. 131–145. Cambridge: M.I.T. Press.

Evans, E.F. 1974*b* Auditory frequency selectivity and the cochlear nerve. In: *Facts and models in hearing* (ed. E. Zwicker & E. Terhard), pp. 118–129. Heidelberg: Springer-Verlag.

Evans, E.F. 1975*a* The cochlear nerve and cochlear nucleus. In *Handbook of sensory physiology*, Vol. V/2 (Auditory System) (ed. W. D. Keidel & W. D. Neff), pp. 1–108. Heidelberg: Springer-Verlag.

Evans, E.F. 1975*b* The sharpening of cochlear frequency selectivity in the normal and abnormal cochlea. *Audiology* **14**, 419–442.

Evans, E.F. 1977 Frequency selectivity at high signal levels of single units in cochlear nerve and nucleus. In *Psychophysics and physiology of hearing* (ed. E. F. Evans & J. P. Wilson), pp. 185–192. London: Academic Press.

Evans, E.F. 1978*a* Place and time coding of frequency in the peripheral auditory system: some physiological pros and cons. *Audiology* **17**, 369–420.

Evans, E.F. 1978*b* Peripheral auditory processing in normal and abnormal ears: physiological considerations for attempts to compensate for auditory deficits by acoustic and electrical prostheses. In *Sensorineural hearing impairment and hearing aids* (ed. C. Ludvigsen & J. Barfod) (*Scand. Audiol. Suppl.* **6**), pp. 9–44.

Evans, E.F. 1981 The dynamic range problem: place and time coding at the level of cochlear nerve and nucleus. In *Neuronal mechanisms of hearing* (ed. J. Syka & L. Aitken), pp. 69–85. New York: Plenum Press.

Evans, E.F. 1982 Functional anatomy of the auditory system. In *The senses* (ed. H. B. Barlow & J. D. Mollon), Chapter 14, pp. 251–306. Cambridge University Press.

Evans, E.F. 1983 Pitch and cochlear fibre temporal discharge patterns. In *Hearing – physiological bases and psychophysics* (ed. R. Klinke & R. Hartmann), pp. 140–145. Berlin, Heidelberg and New York: Springer-Verlag.

Evans, E.F. 1985*a* Aspects of the neuronal coding of time in the mammalian peripheral auditory system relevant to temporal resolution. In *Time resolution in auditory systems: 11th Danavox Symposium* (ed. A. Michelsen), pp. 74–95. Berlin, Heidelberg and New York: Springer Verlag.

Evans, E.F. 1985*b* How to provide speech through an implant device: Dimensions of the problem. In *Cochlear implants* (ed. R. A. Schindler & M. M. Merzenich), pp. 167–184. New York: Raven Press.

Evans, E.F. 1986 Cochlear nerve fibres: temporal discharge patterns, cochlear frequency selectivity and the dominant region for pitch. In *Auditory frequency selectivity* (ed. B. C. J. Moore & R. D. Patterson), pp. 253–264. New York and London: Plenum Press.

Evans, E.F. 1989*a* Cochlear filtering: a view seen through the temporal discharge patterns of single cochlear nerve fibres. In *Cochlear mechanisms - structure function and models* (ed. J. P. Wilson & D. T. Kemp), pp. 241–250. New York: Plenum Press.

Evans, E.F. 1989*b* Lack of correspondence between temporal discharge patterns of cat cochlear nerve fibres and the pitches reported in the Flanagan and Guttman alternate click phase experiment. *Brit. J. Audiol.* **23**, 151.

Evans, E.F. 1989*c* Representation of complex sounds in the peripheral auditory system with particular reference to pitch perception. In *Structure and perception of electroacoustic sound and music* (ed. S. Nielzen & O. Olsson), pp. 117–130. Amsterdam: Excerpta Medica Elsevier.

Evans, E.F. & Harrison, R.V. 1976 Correlation between cochlear outer hair cell damage and deterioration of cochlear nerve tuning properties in the guinea pig. *J. Physiol., Lond.* **256**, 43–44.

Evans, E.F. & Nelson, P.G. 1973 The responses of single neurones in the cochlear nucleus of the cat as a function of their location and the anaesthetic state. *Expl Brain Res.* **17**, 402–427.

Evans, E.F. & Palmer, A.R. 1980 Relationship between the dynamic range of cochlear nerve fibres and their spontaneous activity. *Expl Brain Res.* **40**, 115–118.

Evans, E.F., Pratt, S.R. & Cooper, N.P. 1989 Correspondence between behavioural and physiological frequency selectivity in the guinea pig. *Br. J. Audiol.* **23**, 151–152.

Evans, E.F., Pratt, S.R., Spenner, H. & Cooper, N.P. 1992 Comparisons of physiological and behavioural properties: auditory frequency selectivity. In *Auditory physiology and perception* (ed. Y. Cazals, L. Demany & K. Horner), pp. 159–169. Oxford: Pergamon Press.

Evans, E.F. & Whitfield, I.C. 1964 Classification of unit responses in the auditory cortex of the unanaesthetized and unrestrained cat. *J. Physiol., Lond.* **171**, 476–493.

Evans, E.F. & Wilson, J.P. 1973 Frequency selectivity of the cochlea. In *Basic mechanisms in hearing* (ed. A. R. Moller), pp. 519–551. New York: Academic Press.

Evans, E.F. & Wilson, J.P. 1975 Cochlear tuning properties: Concurrent basilar membrane and single nerve fiber measurements. *Science, Wash.* **190**, 1218–1221.

Evans, E.F. & Zhao, W. 1992*a* Inhibition in the dorsal cochlear nucleus: pharmacological dissection, varieties, nature and possible functions. In *Advances in speech, hearing and language processing*, vol. 3 (*Cochlear nucleus: structure and function in relation to modelling*) (ed. W. A. Ainsworth). London: JAI Press (In the press.)

Evans, E.F. & Zhao, W. 1992*b* Neuropharmacological and neurophysiological dissection of inhibition in the mammalian dorsal cochlear nucleus. In *The mammalian cochlear nuclei: organization and function* (ed. M. Merchan, J. M. Juiz, D. A. Godfrey & E. Magnaini). New York: Plenum Press. (In the Press).

Evans, E.F. & Zhao, W. 1992*c* Varieties of inhibition in the processing and control of processing in the mammalian cochlear nucleus. *Prog. Brain Res.* (In the Press.)

Gold, T. 1948 Hearing II. The physical basis of the action of the cochlea. *Proc. R. Soc. Lond.* B **135**, 492–498.

Harrison, R.V. & Evans, E.F. 1977 The effects of hair cell loss restricted to outer hair cells on the threshold and tuning properties of cochlear fibres in the guinea pig. In *Inner ear biology* (*Coll. I.N.S.E.R.M.* **68**), pp. 105–124.

Harrison, R.V. & Evans, E.F. 1982 Reverse correlation study of cochlear filtering in normal and pathological guinea pig ears. *Hear. Res.* **6**, 303–314.

Khanna, S.M. & Leonard, D.G.B. 1982 Basilar membrane tuning in the cat cochlea. *Science, Wash.* **215**, 305–306.

Liberman, M.C. 1978 Auditory-nerve responses from cats raised in a low noise chamber. *J. acoust. Soc. Am.* **63**, 442–455.

May, B.J. & Sachs, M.B. 1992 Neural encoding of sound intensity in the ventral cochlear nucleus of awake and behaving cats. In *Auditory physiology and perception* (ed. Y. Cazals, L. Demany & K. Horner), pp. 251–262. Oxford: Pergamon Press.

Moller, A.R. 1972 Coding of amplitude and frequency modulated sounds in the cochlear nucleus of the rat. *Acta Physiol. Scand.* **186**, 223–238.

Moller, A.R. 1977 Frequency selectivity of single auditory-nerve fibres in response to broad band noise stimuli. *J. acoust. Soc. Am.* **621**, 135–142.

Patterson, R.D. 1973 The effects of relative phase and the number of components on residue pitch. *J. acoust. Soc. Am.* **53**, 1565–1572.

Palmer, A.R. & Evans, E.F. 1979 On the peripheral coding of the level of individual frequency components of complex sounds at high sound levels. *Expl Brain Res.* **Suppl. II**, 19–26.

Palmer, A.R. & Evans, E.F. 1982 Intensity coding in the auditory periphery of the cat: responses of cochlear nerve and cochlear nucleus neurones to signals in the presence of bandstop noise. *Hear. Res.* **7**, 305–323.

Pick, G.F. & Evans, E.F. 1982 Strategies for high-technology hearing aids to compensate for hearing impairment of cochlear origin. In *High technology aids for disabled people* (ed. W. J. Perkins), Chapter 12, pp. 99–105. London: Butterworths.

Pickles, J.O. 1975 Normal critical bands in the cat. *Acta otolaryngol.* **80**, 245–254.

Sachs, M.B. & Abbas, P.J. 1974 Rate versus level functions for auditory-nerve fibers in cats: tone burst stimuli. *J. acoust. Soc. Am.* **56**, 1835–1847.

Sachs, M.B. & Young, E.D. 1979 Encoding of steady-state vowels in the auditory nerve: Representation in terms of discharge rate. *J. acoust. Soc. Am.* **662**, 470–479.

Schalk, T.B. & Sachs, M.B. 1980 Nonlinearities in auditory nerve fibre responses to band limited noise. *J. acoust. Soc. Am.* **67**(3), 903–913.

Sellick, P.M., Patuzzi, R. & Johnstone, B.M. 1982 Measurements of basilar membrane motion in the guinea-pig using Mossbauer techniques. *J. acoust. Soc. Am.* **72**(1), 131–141.

Suga, N. 1964 Recovery cycles and responses to frequency-modulated tone pulses in auditory neurones of echo-locating bats. *J. Physiol., Lond.* **175**, 50–80.

Viemeister, N.F. 1988 Intensity coding and the dynamic range problem. *Hear. Res.* **34**, 267–274.

West, P.D.B. & Evans, E.F. 1990 Early detection of hearing damage in young listeners resulting from exposure to amplified music. *Br. J. Audiol.* **24**, 89–103.

Whitfield, I.C. & Evans, E.F. 1965 Responses of auditory cortical neurones to stimuli of changing frequency. *J. Neurophysiol.* **28**, 655–672.

Wightman, F.L. & Green, D.M. 1974 The perception of pitch. *Am. Scient.* **62**, 208–215.

Wilson, J.P. & Evans, E.F. 1971 Grating acuity of the ear: psychophysical and neurophysiological measures of frequency resolving power. *Proc. 7th Internat. Cong. Acoustics: Akademiai Kiado, Budapest* **3**, 397–400.

Winter, I.M., Robertson, D. & Yates, G.K. 1990 Diversity of characteristic frequency rate-intensity functions in guinea pig auditory nerve fibres. *Hear. Res.* **45**, 191–202.

Winter, P. & Funkenstein, H. 1971 The auditory cortex of the squirrel monkey: neuronal discharge patterns to auditory stimuli. *Proc. 3rd Cong. Primat. Zurich (1970)*, vol. 2, pp. 24–28. Basel: Kruger.

Young, E.C. 1985 Response characteristics of neurons of

the cochlear nuclei. In *Hearing science* (ed. C. I. Berlin), pp. 423–460. San Diego: College-Hill Press.

Young, E.D. & Sachs, M.B. 1979 Representation of steady-state vowels in the temporal aspects of the discharge patterns of populations of auditory nerve fibres. *J. acoust. Soc. Am.* **665**, 1381–1403.

Zhao, W. & Evans, E.F. 1990 Pharmacological microiontophoretic investigation of receptive-field and temporal properties of units in the cochlear nucleus. *Br. J. Audiol.* **24**, 193.

Basilar membrane responses to two-tone and broadband stimuli

MARIO A. RUGGERO, LUIS ROBLES, NOLA C. RICH AND ALBERTO RECIO

Department of Otolaryngology, University of Minnesota, Research East, 2630 University Avenue S.E., Minneapolis 54414, U.S.A.

SUMMARY

The responses to sound of mammalian cochlear neurons exhibit many nonlinearities, some of which (such as two-tone rate suppression and intermodulation distortion) are highly frequency specific, being strongly tuned to the characteristic frequency (CF) of the neuron. With the goal of establishing the cochlear origin of these auditory-nerve nonlinearities, mechanical responses to clicks and to pairs of tones were studied in relatively healthy chinchilla cochleae at a basal site of the basilar membrane with CF of 8–10 kHz. Responses were also obtained in cochleae in which hair cell receptor potentials were reduced by systemic furosemide injection. Vibrations were recorded using either the Mössbauer technique or laser Doppler-shift velocimetry. Responses to tone pairs contained intermodulation distortion products whose magnitudes as a function of stimulus frequency and intensity were comparable to those of distortion products in cochlear afferent responses. Responses to CF tones could be selectively suppressed by tones with frequency either higher or lower than CF; in most respects, mechanical two-tone suppression resembled rate suppression in cochlear afferents. Responses to clicks displayed a CF-specific compressive nonlinearity, similar to that present in responses to single tones, which could be profoundly and selectively reduced by furosemide. The present findings firmly support the hypothesis that all CF-specific nonlinearities present in the auditory nerve originate in analogous phenomena of basilar membrane vibration. However, because of their lability, it is almost certain that the mechanical nonlinearities themselves originate in outer hair cells.

1. INTRODUCTION

The responses to sound of the mammalian auditory nerve contain many nonlinearities (see review by Ruggero (1992)) whose origins have not been fully specified, largely as a result of the scarcity of data on the representation of sounds in the vibration of the basilar membrane and in the receptor potentials of inner and outer hair cells. One useful distinction among auditory-nerve nonlinearities is whether they are frequency specific: i.e. whether the existence or extent of the nonlinearity is strongly dependent on the spectral content of the stimulus relative to the neuron's characteristic frequency (CF, the frequency to which it is most sensitive). Examples of CF-specific nonlinearities are two-tone rate suppression, two-tone distortion and intensity-dependent phase shifts at near-CF frequencies. Examples of CF-independent nonlinearities are adaptation, reduction of the synchronized responses to one tone by a second tone ('synchrony suppression') and abrupt phase shifts (including 'peak splitting') in responses to tones with frequencies well below CF.

To clarify the cochlear origin of auditory-nerve nonlinearities, we have conducted studies on basilar membrane responses to two-tone and click stimuli. Our findings, reviewed in this paper, demonstrate striking CF-specific mechanical nonlinearities (including two-tone suppression and distortion) and thus strongly support the hypothesis that all frequency-specific auditory-nerve nonlinearities derive from counterparts present in the vibration of the basilar membrane. We also review evidence, based on the effects of furosemide on basilar membrane sensitivity and frequency tuning, that indicates that the frequency specificity of mechanical nonlinearities is not intrinsic to the basilar membrane but rather depends on the integrity of outer hair cell function.

2. METHODS

Mechanical responses to sound were measured from the basilar membrane of the chinchilla at a site located 3.5 mm from the round window. Animals were deeply anesthetized with sodium pentobarbital (initial dose: 65 mg kg^{-1}), tracheotomized and intubated. Normal body temperature was maintained by means of a servo-controlled heating pad. The left pinna was resected, the bulla was widely opened and the tensor tympani muscle was sectioned. A small hole made in the basal turn of the otic capsule allowed direct visualization of the basilar membrane. Basilar membrane vibrations were recorded using either the

Phil. Trans. R. Soc. Lond. B (1992) **336**, 307–315
Printed in Great Britain

307

© 1992 The Royal Society and the authors

[13]

Mössbauer technique (Robles *et al.* 1986) or laser velocimetry (Ruggero & Rich 1991*a*). Both methods, which detect velocity rather than displacement, are based on measurement of the Doppler shift of electromagnetic radiation due to relative motion between a radiation emitter or reflector attached to the basilar membrane and an immobile detector. In the Mössbauer technique, a source of gamma photons (^{57}Fe annealed to a rhodium foil, 6 μm × 80 μm × 80 μm) was placed on the basilar membrane. In the case of laser velocimetry, coherent light from a He–Ne laser was reflected from glass microbeads (10–30 μm).

Acoustic signals (single tones, tone pairs or clicks) were produced under computer control by a custom two-channel waveform generator. The generator's voltage outputs were fed into two Beyer DT-48 earphones mounted on the back of a plastic speculum sealed to the bony ear canal by means of ear-impression compound. A miniature microphone equipped with a probe tube was used to measure the sound pressure within 2 mm of the tympanic membrane.

3. RESULTS

(a) Two-tone suppression

Figure 1 illustrates suppression effects in the responses of a basilar membrane site with CF close to 8 kHz. The velocity-intensity function for an 8-kHz tone presented alone (dashed line) is typical of those obtained in relatively healthy cochleae for tones with frequency close to CF: at very low (and perhaps also at intense) stimulus levels the responses grow linearly but at moderate stimulus intensities responses grow at lower rates (i.e. less than 1 dB dB^{-1}). Upon presentation of a second tone, with frequency lower than CF, the responses to the 8 kHz tone are reduced. In

contrast with 'classical' findings in the auditory nerve (e.g. Javel *et al.* 1978), the mechanical suppression effect varies with probe intensity, being larger at low probe-tone levels. Thus, the input-output functions are not simply shifted to higher probe levels, but are also linearized. Recent findings in the auditory nerve indicate that, in fact, similar suppression-induced changes in rate-intensity functions can be demonstrated in the responses of high-threshold, low-spontaneous-rate cochlear afferents (Sokolowski *et al.* 1989).

As the intensity of the suppressor tone is raised, the magnitude of the mechanical suppression effect increases monotonically (figure 1). The slope of the function relating suppression to suppressor intensity depends on frequency of the suppressor tone relative to CF, both in the basilar membrane (Ruggero *et al.* 1992) and in the cochlear nerve (Costalupes *et al.* 1987; Delgutte 1990). For suppressors with frequency higher than CF, the rate is low, in the order of 0.4 dB dB^{-1}; for suppressors with frequency well below CF (as illustrated in figure 1) the rate is substantially higher (between 0.65 and 1.42 dB dB^{-1}).

Figure 2 shows that two-tone suppression in the basilar membrane, much as two-tone rate suppression in the auditory nerve (e.g. Abbas 1978; Schmiedt 1982), depends on probe frequency, being largest when probe frequency equals CF (10 kHz) and non-existent for probe frequencies far removed from CF. The CF specificity of mechanical suppression also holds for suppressor tones with frequencies below CF (not shown). In this and several other respects, including its sensitivity to cochlear damage, two-tone rate suppression has mechanical counterparts in the basilar

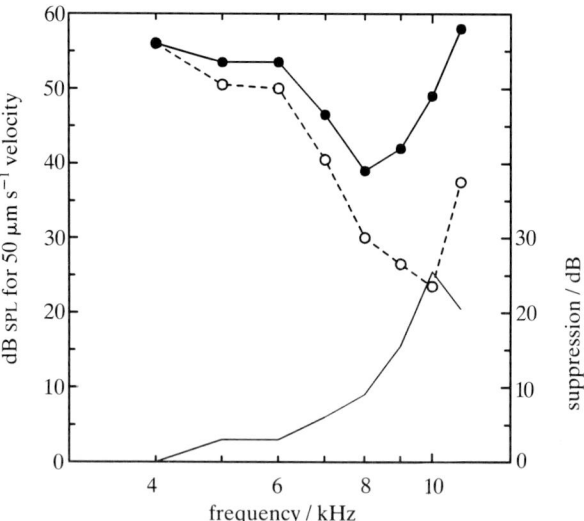

Figure 2. Changes in basilar-membrane frequency tuning produced by a 12 kHz suppressor tone presented at 63 dB SPL. The isovelocity tuning curves (scale in left ordinate) were derived from intensity functions for probe tones presented alone (open circles, dashed line) and in the presence of the suppressor tone (filled circles, solid line). The thin solid line indicates suppression magnitude (scale in right ordinate) as a function of probe frequency. The 12 kHz suppressor tone evoked a response 7.4 dB smaller than the 50 μm s^{-1} isovelocity criterion. Cochlea: L14.

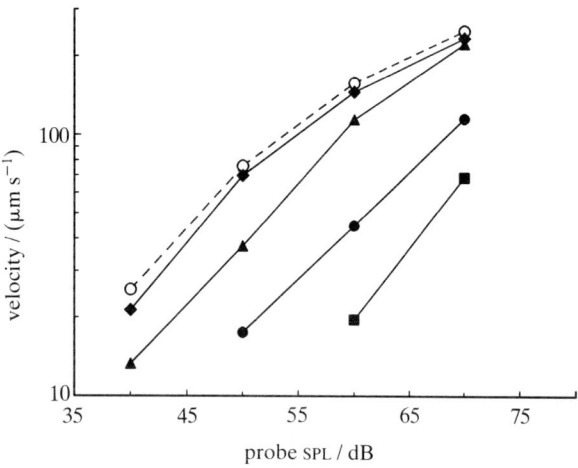

Figure 1. Effect of suppressor tones on basilar membrane intensity functions for a near-CF tone. Intensity functions are shown for an 8 kHz probe tone alone (open circles, dashed line) and in the presence of a 300 Hz suppressor at several intensities (filled symbols, solid lines): diamonds, 70 dB; triangles, 80 dB; circles, 90 dB; squares, 100 dB. Cochlea: L29.

membrane and thus probably derives from these counterparts (Ruggero *et al.* 1992).

(b) *Two-tone distortion*

When listening to pairs of tones, humans hear additional tones – distortion products or combination tones – that are not present in the stimulus. Substantial psychophysical (e.g. Goldstein 1967; Smoorenburg 1972), neurophysiological (e.g. Siegel *et al* 1982) and acoustic (e.g. Mountain 1980) evidence suggests that combination tones arise on the basilar membrane. However, with the possible exception of an f_2–f_1 vibration component reported by Rhode (1977), searches for mechanical combination tones in the basilar membrane were until recently unsuccessful (Wilson & Johnstone, 1973; Rhode, 1977), perhaps due to a combination of inadequate vibration-measurement technologies and the poor physiological state of the experimental cochleae. During the last year mechanical combination tones have been finally measured, using laser velocimetry, in the cochleae of guinea pig and chinchilla (Nuttall *et al.* 1990; Robles *et al.* 1990, 1991).

Figure 3 depicts the frequency spectrum of the mechanical response to a pair of tones, recorded from a basal site of the chinchilla basilar membrane. The tones were chosen with widely spaced frequencies (13.33 and 16.66 kHz) such that the cubic difference tone ($2f_1$–f_2) coincided with the CF (10 kHz) of the basilar membrane site. At high stimulus intensities the spectrum has components with frequencies corresponding to those of both primary tones and to the cubic difference tone $2f_1$–f_2. However, at the lowest stimulus intensities (60 and 70 dB SPL) the spectrum

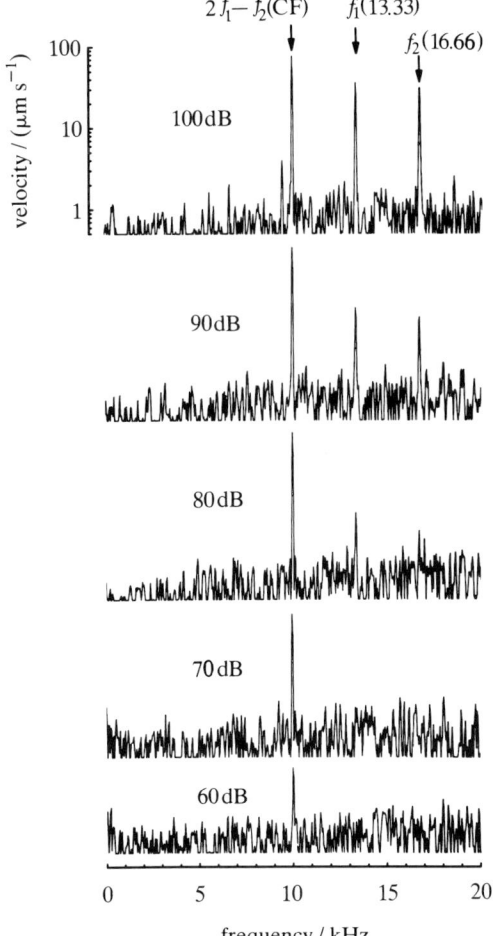

Figure 3. Frequency spectra of basilar membrane responses to a two-tone stimulus. The frequencies of equal-intensity tones (f_1, 13.33 kHz; f_2, 16.66 kHz) were chosen so that $2f_1$–f_2 coincided with the CF of the basilar membrane site (10 kHz). Stimulus intensity (in dB relative to 20 μPa) is indicated in each panel. Cochlea: L47.

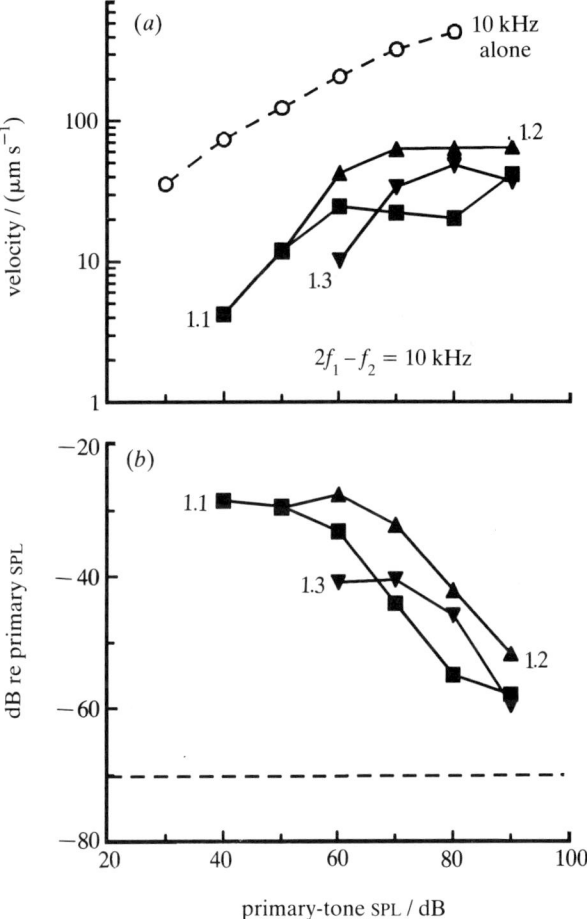

Figure 4. Magnitudes of $2f_1$–f_2 distortion products as a function of primary-tone intensity and frequency ratio (solid lines). The frequencies of the two primary tones were chosen so that $2f_1$–f_2 coincided with CF (10 kHz). (a) Open circles and dash line represent the velocity input–output function for a single tone at the distortion-product frequency. Closed symbols and solid lines indicate input–output functions for pairs of tones with f_2:f_1 ratios of 1.1, 1.2 and 1.3. (b) Distortion-product magnitudes are plotted as effective level, i.e. as intensity of a single tone at the distortion-product frequency required to produce a response of the same magnitude as the distortion product produced by the two-tone stimulus. Effective levels are expressed as decibels relative to primary-tone intensities. The dash line indicates that artifactual $2f_1$–f_2 distortion products in the acoustic-stimulus system were about 70 dB less than the intensity of the primary tones. Cochlea: L47.

[15]

contains only one detectable component, the 10 kHz combination tone. The absence of detectable responses at the frequencies of the primary tones is due to the sharp filtering imposed by the cochlear partition on frequencies higher than CF. The fact that the responses to the primary tones are substantially lower than to the $2f_1$–f_2 combination tone implies that the latter arose at a remote basal site, presumably one with CF intermediate between the frequencies of the primary tones, and subsequently propagated apically to the recording site with CF of 10 kHz.

Figure 4 (top) shows distortion data obtained in the same cochlea from which the spectrum of figure 3 was recorded. The dashed line depicts a velocity-intensity function for responses to a CF tone. The other lines indicate velocities at $2f_1 - f_2$ for responses to primary tones with frequencies f_1 and f_2 chosen such that $2f_1$–f_2 coincided with CF, and f_2:f_1 ratios were 1.1, 1.2 and 1.3. In general, the curves grow monotonically with stimulus intensity but saturate at levels of 60–80 dB SPL. To permit comparison with neurophysiological and psychophysical findings, the data are replotted in the bottom panel of figure 4 with the magnitude of the distortion product being expressed as effective level: intensity of a single tone at the distortion-product frequency (in this case CF) required to produce a response of the same magnitude as the distortion product due to the two-tone stimulus. As previously shown for distortion products in the auditory nerve (Buunen & Rhode 1978) effective levels of the mechanical distortion products are largest at the lowest stimulus intensities and decline monotonically as intensity is raised. Even at stimulus levels as high as 90 dB SPL the measured distortion products are larger than artifactual distortion products in the acoustic-stimulus system (dashed line), indicating that the distortion products originate in the cochlea. The largest magnitudes of basilar membrane distortion products are comparable with those measured neurophysiologically. Unexpectedly, in figure 4 there is no clear monotonic reduction of distortion product magnitude as the f_2:f_1 ratio is increased from 1.1 to 1.3; such a monotonic reduction, which has been seen in recordings from guinea pig basilar membrane (Nuttall *et al.* 1990) and is a common (but not universal) feature of two-tone distortion measured from cochlear afferents (Buunen & Rhode, 1978), may have been absent in cochlea L47 due to organ of Corti damage at regions located basal to the recording site or may reflect a complex interaction between two-tone distortion and suppression.

(c) *Responses to clicks*

Clicks have been previously used as stimuli in a single experimental series including relatively healthy cochleae with sharply tuned and nonlinear responses (Robles *et al.* 1976). Although those recordings were limited by the severely nonlinear input–output characteristic of the Mössbauer technique, which rectified and clipped the click responses, the findings were straightforward: they confirmed the existence of a frequency specific nonlinearity in the squirrel

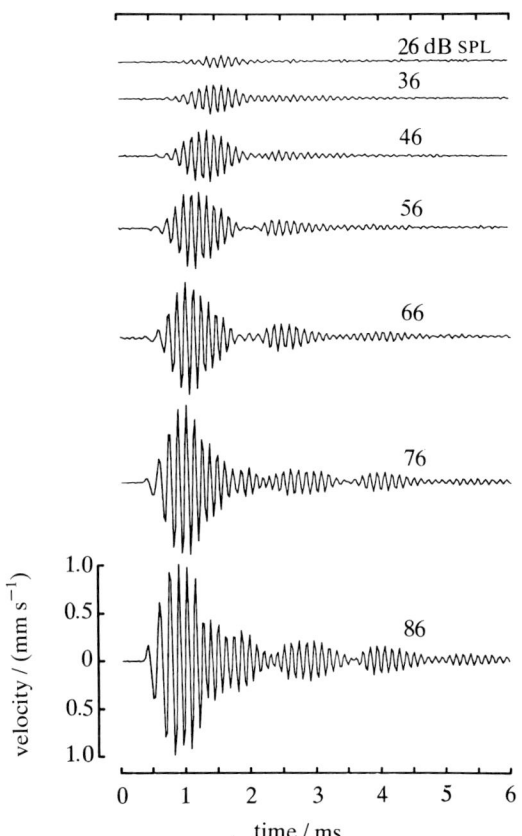

Figure 5. Velocity responses to rarefaction clicks of a basilar membrane site with CF of 9 kHz. Click intensity, expressed as peak pressure (in decibels relative to 20 μPa), is indicated for each tracing. Modified from Ruggero & Rich (1991a). Cochlea: L13.

monkey basilar membrane and demonstrated time-domain counterparts of certain features of responses to tones (e.g. Rhode 1971). Taking advantage of a newly-developed method, laser vibrometry, which is essentially linear and much more sensitive than the Mössbauer technique (Ruggero & Rich 1991a), we now are able to extend the findings of the Mössbauer study.

Figure 5 shows velocity responses to clicks of a basal site of a basilar membrane recorded in an exceptionally undamaged chinchilla cochlea. Responses to low-level clicks consist of high-frequency vibrations modulated by a spindle-shape envelope and are reminiscent of 'revcors' (reverse correlations) for low-CF auditory-nerve responses to noise (see, for example, de Boer & de Jongh 1978). With increases in click intensity, the responses grow monotonically but at rates lower than linear; further, different segments of the response grow at different rates, so that the responses at different click intensities are not scaled versions of each other. In general, the centre of gravity shifts to earlier times as intensity is raised. Although not obvious in figure 5, at the highest intensities the earliest vibration cycle (but no others) grow linearly, thus establishing an absolute, irreducible latency. This latency, which when measured from the onset of stapes displacement amounts to only 90 μs, presumably corresponds to the time that it takes the traveling wave to move from the oval window to the recording site 3.5 mm away (Ruggero *et al.* 1991a).

Phil. Trans. R. Soc. Lond. B (1992)

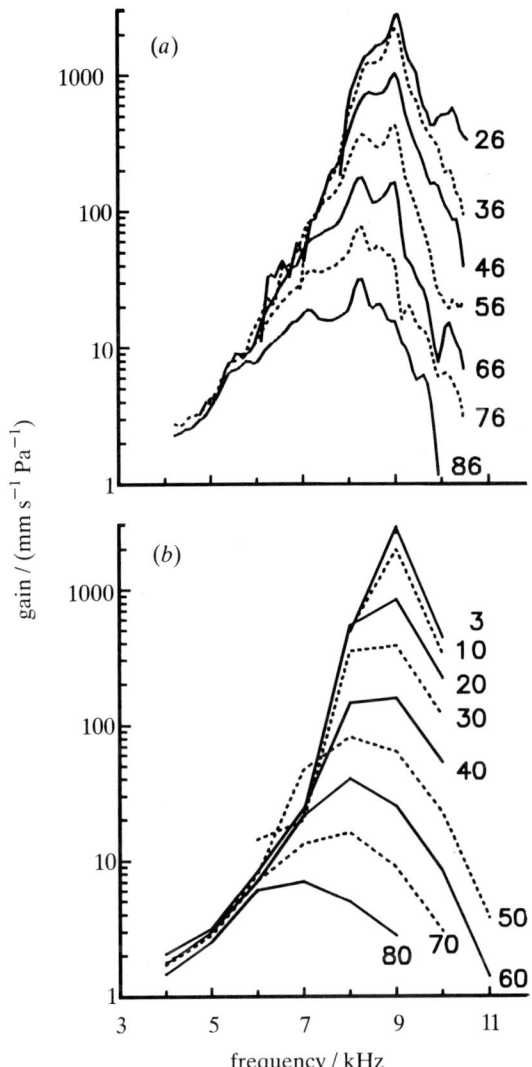

Figure 6. Gain-versus-frequency functions for responses to (*a*) clicks and (*b*) tones of a single basilar membrane site. Gains were computed by dividing, frequency by frequency, spectral response magnitude by stimulus pressure. The spectral magnitudes for click responses were obtained by Fourier transformation of the data of figure 5. Click peak pressures are indicated next to each gain spectrum. Cochlea: L13.

Figure 6 (top) shows gain-versus-frequency functions computed from the click responses of figure 5. In accordance with the compressive nonlinearity evident in the time-domain representations, the gain functions change systematically as a function of click level. Peak gains are largest (about 3 m s^{-1} Pa^{-1}) at CF (9 kHz) at the lowest stimulus intensities; as click intensities are raised, the peak spectral velocities shift to lower frequencies and are reduced in magnitude. For a thousand-fold (60 dB) increase in click pressure, there is a hundred-fold (40 dB) reduction in peak gain. At frequencies well below CF the gain functions superimpose, indicating that at these frequencies responses grow linearly. For comparison with the click data, gain functions for responses to tones are plotted in the bottom panel of figure 6 using scales identical to those in the top panel. It is apparent that the two sets of

gain functions are very similar. Thus, for example, responses to 26 dB clicks and 3 dB tones are nearly linear and yield nearly identical peak gains (about 3 m s^{-1} Pa^{-1}); similarly, responses to stimuli about 1000-fold more intense (86 dB clicks and 60 dB tones) yield peak gains of 30–40 mm s^{-1} Pa^{-1}. Thus, the CF specific compressive nonlinearities evident in responses to tones or clicks appear to be the same and do not preclude predicting the responses to one type of stimulus from these evoked by the other.

(*d*) *Lability of CF-specific nonlinear mechanical responses*

Twenty years ago Rhode demonstrated that a CF-specific compressive nonlinearity existed in the response of the squirrel monkey basilar membrane (Rhode 1971) and that the nonlinearity disappeared upon death (Rhode 1973). With the advantage of hindsight, it now seems certain that failures to confirm the existence of CF-specific nonlinearities in basilar membrane responses during the next decade were due to the poor state of the experimental cochleae. More recent experiments (Sellick *et al.* 1982; Robles *et al.* 1986) suggest that basilar membrane nonlinearities are inextricably related to frequency tuning and that the same processes that abolish nonlinearities also destroy sharp frequency tuning. Thus, to the extent that nonlinearities in responses to two-tone or broadband stimuli are CF specific, one should expect them also to be labile. This indeed appears to be the case.

Cochleae that are relatively healthy, and thus yield sharply frequency-tuned and sensitive responses to tones, sustain strong two-tone mechanical suppression effects; conversely, poorly tuned and insensitive cochleae produce weak suppression effects (Ruggero *et al.* 1992). The lability of mechanical suppression is consistent with electrophysiological recordings from the auditory nerve: afferent fibers innervating cochlear regions devoid of outer hair cells do not exhibit rate suppression (see Schmiedt & Zwislocki 1980). Although we have not yet explored the lability of mechanical distortion-product generation in a systematic way, the absence of distortion products in insensitive cochleae and their presence in healthy cochleae suggests that the lability of two-tone distortion already demonstrated in auditory-nerve recordings (e.g. Siegel *et al.* 1982) must have correlates in the basilar membrane. Finally, the CF-specific nonlinearity and sensitivity of responses to clicks disappear after death, paralleling similar changes in responses to tones (Ruggero & Rich 1991*a*; Ruggero *et al.* 1991*a*).

The lability of basilar membrane sensitivity and CF-specific nonlinearities suggests that their origin does not reside in the largely acellular, and thus probably metabolically-insensitive, basilar membrane. Rather, there is a consensus belief that such basilar membrane properties reflect the activity of the living cells of the organ of Corti, probably the outer hair cells. Although substantial, evidence for this hypothesis has been mostly indirect. The strongest indirect evidence is the alteration of combination-tone otoacoustic emissions by electrical stimulation of the medial efferent system

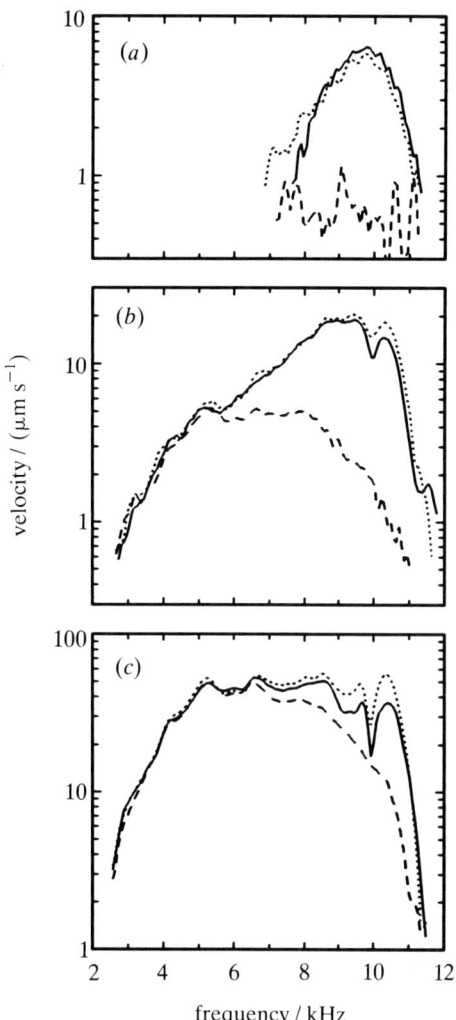

Figure 7. Frequency-specific effects of furosemide upon the magnitude of basilar membrane responses to clicks. The velocity-magnitude frequency spectra were computed by Fourier transformation of responses to clicks with peak pressures of (*a*) 48, (*b*) 68 and (*c*) 88 dB (relative to 20 μPa). For each stimulus level, three curves are displayed, representing responses immediately preceding (solid line, −13 min) and following (long-dash line, +9 min) a furosemide injection and when full recovery had occurred (short-dash line, +100 min). Redrawn from Ruggero & Rich (1991*b*). Cochleae: L14.

(Mountain 1980; Siegel & Kim 1982). We have recently taken advantage of a relatively specific and reversible cochlear manipulation, namely furosemide-induced reductions of receptor potentials, to test the hypothesis that the receptor potentials of outer hair cells control the mechanical frequency selectivity and sensitivity of basilar membrane vibrations (Ruggero & Rich 1991*b*).

Figure 7 shows alterations of responses to clicks caused by intravenous injections of furosemide, a loop-inhibiting diuretic that rapidly and reversibly reduces the hair cell receptor potentials. The spectra for responses to 48 dB and 68 dB clicks are drastically altered by furosemide: sensitivity drops by at least 20 dB at CF but it is not changed at frequencies well below CF. In a manner qualitatively similar to the effects

of two-tone suppression (see figure 2) the sharpness of frequency tuning is reduced and the frequency of maximal sensitivity shifts to lower frequencies. Also similar to the effects of suppression, the effects of furosemide depend on stimulus intensity: they are largest for low-level clicks and small for intense clicks.

We interpret the mechanical effects of furosemide as showing that the organ of Corti determines basilar membrane vibrations. More specifically, we argue, these effects almost inescapably imply that the receptor potentials of the outer hair cells participate in a feedback control system that normally boosts the vibration magnitude of the basilar membrane. Our argument is based on the following facts (reviewed by Ruggero & Rich (1991*b*)): (i) receptor potentials are produced by inner and outer hair cells, but not by supporting cells; (ii) furosemide alters the receptor potentials of hair cells either directly (unlikely), or indirectly (most likely) via reduction of the endocochlear potential; (iii) inner hair cells are unlikely to link furosemide-induced changes in receptor potentials to a mechanical effect: the effects of injecting negative extracellular DC current into scala media, a procedure (similar to reducing the endocochlear potential) which causes reduction of auditory-nerve and inner hair cell responses, cannot be mimicked by passing currents into inner hair cells via intracellular electrodes (Nuttall 1985). Furthermore, only outer hair cells display electrically induced motility (see Brownell *et al.* 1985) and they (but not inner hair cells) are positioned at sites favourable for influencing basilar membrane mechanics (see Lim 1980).

4. DISCUSSION

Our studies on the mechanical responses of the chinchilla basilar membrane to two-tone and broad-band stimuli lead us to conclude that the main features of the CF-specific nonlinearities evoked by these stimuli in the auditory nerve derive from similar features of basilar membrane vibration. This conclusion is consistent with work of other laboratories on basilar-membrane responses to tones and clicks in the cochleae of squirrel monkeys (Rhode 1971; Robles *et al.* 1976) and guinea pigs (Sellick *et al.* 1982), on the low-frequency biasing of guinea pig basilar membrane responses to CF stimuli (Patuzzi *et al.* 1984) and on two-tone distortion at the guinea pig basilar membrane (Nuttall *et al.* 1990). There is a high likelihood that all CF-specific auditory-nerve nonlinearities have correlates in basilar membrane motion. The effects of furosemide on basilar membrane sensitivity and tuning imply that, in turn, the mechanical nonlinearities reflect an influence of the outer hair cells. The big unanswered question, of course, is whether the mild intrinsic frequency tuning of the basilar membrane (e.g. as measured during furosemide intoxication) is augmented by its interaction with the outer hair cells (and, indirectly, the tectorial membrane) or whether, alternatively, outer hair cells are intrinsically sharply tuned (e.g. Brundin *et al.* 1989) and thus responsible for sharp basilar membrane tuning in normal cochleae.

Having specified the basilar membrane – outer hair cell complex as the origin of CF-specific nonlinearities, it would seem that most other types of nonlinearities must arise in inner hair cells or at their synapses with cochlear afferent terminals. Compressive and rectifying nonlinearities exist in inner hair cells, at the process that transduces stereociliar displacement into electrical potentials (Russell *et al.* 1986), and at the synapses between inner hair cells and afferent terminals of the auditory nerve (see Greenwood, 1986). Because certain features of two-tone suppression can be generated by any process which exhibits compressive or rectifying properties (see Geisler, 1985) these two sites probably participate in generating 'synchrony suppression'. An inner hair cell or synaptic origin of 'synchrony suppression' would also be consistent with its apparent insensitivity to outer hair cell loss (Javel *et al.* 1983). The inner hair cell compressive nonlinearity should also account for the nonlinear growth of auditory nerve responses to tones with frequency well below CF. On the other hand, the absence of adaptation in either basilar membrane or inner hair cell responses suggests that adaptation must arise at the chemical synapses between inner hair cells and type I cochlear afferents.

One type of auditory-nerve nonlinearity whose origin remains mysterious is the nearly universal intensity-dependent phase shift of responses to low-frequency tones: 'peak splitting', the appearance of two preferred times of phase locking and its slightly less-striking version, abrupt shifts in response phases (e.g. Kiang & Moxon 1972; Gifford & Guinan 1983; Ruggero & Rich 1989). We have obtained mechanical data ruling out a basilar membrane origin for these phenomena (Ruggero *et al.* 1991*b*). This is as expected on the basis of previously demonstrated linearity of responses to stimuli with frequency well below CF. What is puzzling is that clear counterparts of abrupt intensity-dependent phase shifts have been demonstrated in inner hair cells only rarely (Cody & Mountain 1989; Dallos & Cheatham 1989).

This investigation was financially supported principally by NIH (NIDCD) grants DC-00110 and DC-00419.

REFERENCES

Brownell, W.E., Bader, C.R., Bertrand, D. & de Ribaupierre, Y. 1985 Evoked mechanical responses of isolated outer hair cells. *Science, Wash.* **227**, 194–196.

Brundin, L., Flock, A. & Canlon, B. 1989 Sound-induced motility of isolated cochlear outer hair cells is frequency-specific. *Nature, Lond.* **342**, 814–816.

Buunen, T.J.F. & Rhode, W.S. 1978 Responses of fibers in the cat's auditory nerve to the cubic difference tone. *J. acoust. Soc. Am.* **64**, 772–781.

Cody, A.R. & Mountain, D.C. 1989 Low-frequency responses of inner hair cells: evidence for a mechanical origin of peak splitting. *Hear. Res.* **41**, 89–100.

Costalupes, J.A., Rich, N.C. & Ruggero, M.A. 1987 Effects of excitatory and non-excitatory suppressor tones on two-tone rate suppression in auditory nerve fibers. *Hear. Res.* **26**, 155–164.

Dallos, P. & Cheatham, M.A. 1989 Nonlinearities in

cochlear receptor potentials and their origins. *J. acoust. Soc. Am.* **86**, 1790–1796.

de Boer, E. & de Jongh, H.R. 1978 On cochlear encoding: potentialities and limitations of the reverse-correlation technique. *J. acoust. Soc. Am.* **63**, 115–135.

Delgutte, B. 1990 Two-tone rate suppression in auditory-nerve fibers: dependence on suppressor frequency and level. *Hear. Res.* **49**, 225–246.

Geisler, C.D. 1985 Effects of a compressive nonlinearity in a cochlear model. *J. acoust. Soc. Am.* **78**, 257–260.

Gifford, M.L. & Guinan, J.J. Jr 1983 Effects of crossed-olivocochlear-bundle stimulation on cat auditory nerve fiber responses to tones. *J. acoust. Soc. Am.* **74**, 115–123.

Goldstein, J.L. 1967 Auditory nonlinearity. *J. acoust. Soc. Am.* **41**, 676–689.

Greenwood, D.D. 1986 What is 'synchrony suppression'? *J. acoust. Soc. Am.* **79**, 1857–1872.

Javel, E., Geisler, C.D. & Ravindran, A. 1978 Two-tone suppression in auditory nerve of the cat: rate-intensity and temporal analysis. *J. acoust. Soc. Am.* **63**, 1093–1104.

Javel, E., McGee, J.A., Walsh, E.J. & Farley, G.R. 1983 Studies of 'synchrony suppression' in normal and hearing-impaired cats. In *Mechanics of hearing* (ed. W. R. Webster & L. M. Aitkin), pp. 46–51. Clayton, Victoria, Australia. Monash University Press.

Kiang, N.Y.S. & Moxon, E.C. 1972 Physiological considerations in artificial stimulation of the inner ear. *Ann. Otol. Rhinol. Laryngol.* **81**, 714–730.

Lim, D.J. 1980 Cochlear anatomy related to cochlear micromechanics. A review. *J. acoust. Soc. Am.* **67**, 1686–1695.

Mountain, D.C. 1980 Changes in endolymphatic potential and crossed olivocochlear bundle alter cochlear mechanics. *Science, Wash.* **210**, 71–72.

Nuttall, A.L. 1985 Influence of direct current on DC receptor potentials from cochlear inner hair cells in the guinea pig. *J. acoust. Soc. Am.* **77**, 165–175.

Nuttall, A.L., Dolan, D.F. & Avinash, G. 1990 Measurements of basilar membrane tuning and distortion with laser Doppler velocimetry. In *The mechanics and biophysics of hearing*, (ed. P. Dallos, C. D. Geisler, J. W. Matthews, M. A. Ruggero & C. R. Steele), pp. 288–295. Berlin: Springer-Verlag.

Patuzzi, R., Sellick, P.M. & Johnstone, B.M. 1984 The modulation of the sensitivity of the mammalian cochlea by low frequency tones. III. Basilar membrane motion. *Hear. Res.* **13**, 19–27.

Rhode, W.S. 1971 Observations of the vibration of the basilar membrane in squirrel monkeys using the Mössbauer technique. *J. acoust. Soc. Am,* **49,** 1218–1231.

Rhode, W.S. 1973 An investigation of post-mortem cochlear mechanics using the Mössbauer effect. In *Basic mechanisms in hearing* (ed. A. R. Møller), pp. 49–63. New York: Academic Press.

Rhode, W.S. 1977 Some observations on two-tone interaction measured with the Mössbauer effect. In *Psychophysics and physiology of hearing* (ed. E. F. Evans & J. P. Wilson), pp. 27–41. London: Academic Press.

Robles, L., Rhode, W.S. & Geisler, C.D. 1976 Transient response of the basilar membrane measured in the squirrel monkey using the Mössbauer effect. *J. acoust. Soc. Am.* **59**, 926–939.

Robles, L., Ruggero, M.A. & Rich, N.C. 1986 Basilar membrane mechanics at the base of the chinchilla cochlea. I. Input-output functions, tuning curves, and response phases. *J. acoust. Soc. Am.* **80**, 1364–1374.

Robles, L., Ruggero, M.A. & Rich, N.C. 1990 Two-tone distortion products in the basilar membrane of the chinchilla cochlea. In *The mechanics and biophysics of hearing* (ed. P. Dallos, C. D. Geisler, J. W. Matthews, M. A. Ruggero & C. R. Steele), pp. 304–311. Berlin: Springer-Verlag.

Robles, L., Ruggero, M.A. & Rich, N.C. 1991 Two-tone distortion in the basilar membrane of the cochlea. *Nature, Lond.* **349**, 413–414.

Ruggero, M.A. 1992 Physiology and coding of sound in the auditory nerve. In *The mammalian auditory pathway: neurophysiology* (ed. R. Fay & A. N. Popper). New York: Springer-Verlag. (In the press.)

Ruggero, M.A. & Rich, N.C. 1989 'Peak splitting': intensity effects in cochlear afferent responses to low-frequency tones. In *Cochlear mechanisms–structure, function and models* (ed. J. P. Wilson & D. T. Kemp), pp. 259–266. London: Plenum.

Ruggero, M.A. & Rich, N.C. 1991a Application of a commercially-manufactured Doppler-shift laser velocimeter to the measurement of basilar-membrane vibration. *Hear. Res.* **51**, 215–230.

Ruggero, M.A. & Rich, N.C. 1991b Furosemide alters organ of Corti mechanics: evidence for feedback of outer hair cells upon the basilar membrane. *J. Neurosci.* **11**, 1057–1067.

Ruggero, M.A., Rich, N.C. & Recio, A. 1991a Basilar membrane responses to clicks. In *Auditory physiology and perception* (ed. Y. Cazals, L. Demany & K. Horner), pp. 85–91. Oxford: Pergamon Press.

Ruggero, M.A., Rich, N.C. & Robles, L. 1991b Comparison of cochlear-nerve and basilar-membrane responses to low-frequency tones: absence of macromechanical basis for 'peak splitting'. *Assoc. Res. Otolaryngol. Midwinter Meet. Abst.* **14**, 78.

Ruggero, M.A., Robles, L. & Rich, N.C. 1992 Two-tone suppression in the basilar membrane of the chinchilla cochlea: mechanical basis of auditory-nerve two-tone rate suppression. *J. Neurophysiol.* (Submitted.)

Russell, I.J., Richardson, G.P. & Cody, A.R. 1986 Mechanosensitivity of mammalian auditory hair cells *in vitro*. *Nature, Lond.* **321**, 517–519.

Schmiedt, R.A. & Zwislocki, J.J. 1980 Effects of hair cell lesions on responses of cochlear nerve fibers. II. Single- and two-tone intensity functions in relation to tuning curves. *J. Neurophysiol.* **43**, 1390–1405,

Sellick, P.M., Patuzzi, R. & Johnstone, B.M. 1982 Measurement of basilar membrane motion in the guinea pig using the Mössbauer technique. *J. acoust. Soc. Am.* **72**, 131–141.

Siegel, J.H. & Kim, D.O. 1982 Efferent neural control of cochlear mechanics? Olivocochlear bundle stimulation affects cochlear biomechanical nonlinearity. *Hear. Res.* **6**, 171–182.

Siegel, J.H., Kim, D.O. & Molnar, C.E. 1982 Effects of altering organ of Corti on cochlear distortion products f_2-f_1 and $2f_1-f_2$. *J. Neurophysiol.* **47**, 303–328.

Smoorenburg, G.F. 1972 Combination tones and their origin. *J. acoust. Soc. Am.* **52**, 615–632.

Sokolowski, B.H.A., Sachs, M.B. & Goldstein, J.L. 1989 Auditory nerve rate-level functions for two-tone stimuli: possible relation to basilar membrane nonlinearity. *Hear. Res.* **41**, 115–124.

Wilson, J.P. & Johnston, J.R. 1973 Basilar membrane correlates of the combination tone $2f_1-f_2$. *Nature, Lond.* **241**, 206–207.

Discussion

A. M. Brown (*Laboratory of Experimental Psychology, University of Sussex, U.K.*). We have also measured distortion from the cochlea, but using the far less sophisticated technique of recording from a microphone in the ear canal. Professor Ruggero's results are very interesting because they show a different picture from the one that we see. We think that the difference between basilar membrane and ear canal recorded responses may be crucial to our understanding of how the cochlea achieves its frequency selectivity.

At low stimulus levels, we see relatively little re-emission of stimulus energy, but plenty of distortion which reaches a maximum when its frequency is just over half a octave below the high frequency tone (f_2). We have concluded that there is a broadly tuned resonance at the f_2 site which is tuned to a frequency half an octave below f_2 (Brown & Gaskill 1990; A. M. Brown, S. A. Gaskill & D. M. Williams, unpublished results, 1991). We suspect that this may be due to the mechanical properties of the tectorial membrane (TM). Our data may offer support for the two-resonance model of Allen (1980), whereby the TM would effectively sharpen the tuning of the stimulus to inner hair cells by providing an alternative path for vibrational energy immediately below the characteristic frequency (CF) to reach the cochlear fluids, bypassing the subtectorial space. From this we can suggest that the distortion matnitude 'profile' as a function of frequency reflects the frequency response of the tectorial membrane at that place in the cochlea, when this is measured in the ear canal. But the distortion measured in the basilar membrane response may differ greatly depending on the extent to which TM resonance influences that of the BM.

The response Professor Ruggero has measured from the basilar membrane when stimulating with two tones shows distortion components that increase in magnitude (rather than decreasing like ours) as they approach the stimulus frequency. If our thinking is correct, his recordings of the basilar membrane response near the f_2 place may reveal a relative minimum in distortion level (especially for low stimulus levels) when the distortion frequency is half an octave below f_2. Do they?

References

Allen, J.B. 1980 Cochlear micromechanics – A physical model of transduction. *J. acoust. Soc. Am.* **68**, 1660–1670.

Brown, A.M. & Gaskill, S.A. 1990 Can basilar membrane tuning be inferred from distortion measurement? In *The mechanics and biophysics of hearing* (ed. P. Dallos, C. D. Geisler, J. W. Matthews, M. A. Ruggero & C. R. Steele), pp. 164–169. Berlin: Springer Verlag.

M. A. Ruggero. Regarding the variation of distortion magnitude with $2f_1-f_2$ at CF and variable $f_2:f_1$ ratio: we are not yet certain whether in the chinchilla basilar membrane there is a monotonic decrease of distortion magnitude as a function of increasing separation between the primaries. Working in the guinea pig basilar membrane, Nuttall *et al.* (1990) have measured a seemingly monotonic decrease at a rate of 100–200 dB per octave. On the other hand, in their auditory-nerve study Buunen & Rhode (1978) often found nonmonotonic functions.

Answering Dr Brown's question concerning a

possible distortion-magnitude minimum at a frequency half an octave below f_2 would require a stimulus protocol with f_2 at CF and a variable f_1. We have not yet used such a protocol in our basilar membrane experiments.

E. F. EVANS (*Department of Communication and Neuroscience, University of Keele, U.K.*). I do not believe that there is such a clear distinction as you suggest between tone and noise stimuli in producing nonlinear responses in cochlear nerve fibres. Although nonlinear effects are obvious with two-tone stimuli, they become rapidly less obvious with the addition of more frequency components. The experiments I described (Evans, this symposium; see also Evans. 1989) on the relative invariance of cochlear fibre tuning as seen through the weighting of temporal discharge patterns with stimulus level, occurred with harmonic complexes (i.e. multiple tones) as with click series and noise stimuli (i.e. broadband).

Even the apparently gross nonlinear effects described by Horst *et al.* (1985) can be accounted for by a model (Evans 1980) having linear tuning followed by an automatic gain control having a time constant small enough relative to the period of the waveform generated by their very low fundamental stimuli. In fact, linear filtering with a linear time window centred on the portion of the waveform between the waveform peaks is enough to show these anomalous effects (Evans 1987, 1988).

References

Evans, E.F. 1980 An electronic analogue of single unit recording from the cochlear nerve for teaching and research. *J. Physiol., Lond.* **298**, 6–7.

Evans, E.F. 1987 Modelling cochlear nerve fibre responses to complex pitch-producing stimuli. *Br. J. Audiol.* **21**, 311.

Evans, E.F. 1988 Cochlear nerve discharge patterns in response to complex stimuli: model predictions and neural data. *Br. J. Audiol.* **22**, 136.

Evans, E.F. 1989 Cochlear filtering: a view seen through the temporal discharge patterns of single cochlear nerve fibres. In *Cochlear mechanisms – structure function and models* (ed. J. P. Wilson & D. T. Kemp), pp. 241–250. New York: Plenum.

Horst, J.W., Javel, E. & Farley, G.R. 1985 Extraction and enhancement of spectral structure by the cochlea. *J. acoust. Soc. Am.* **78**, 1898–1901.

M. A. RUGGERO. Professor Evans' comment appears to refer to my verbal presentation rather than to the paper. I agree that there is no clear-cut distinction between tones and broadband stimuli in their ability to evoke nonlinearities in cochlear nerve fibers (or the basilar membrane). During my presentation I merely attempted to contrast (i) the clear nonlinearities in responses to two-tones (distortion and suppression) with (ii) the fact that the nonlinearities in basilar membrane responses to both clicks and single tones do not preclude predicting reasonably accurately the responses to one type of stimulus from those to the other.

E. F. EVANS. The concept of positive-feedback of mechanical energy to enhance the sensitivity and selectivity of the cochlea, while compelling, has several problems. At least two questions arise.

1. If the feedback occurs from the outer hair cells to the inner hair cells via local structures (e.g. the tectorial membrane), how far should the sharp tuning be reflected in basilar membrane motion? In other words, how much mechanical coupling should we expect between the mechanics of the apical end of the hair cells and the basilar membrane? Could some, at least, of the variation in reported mechanical tuning be ascribed to variable degrees of such coupling?

2. A simple positive-feedback mechanism for enhancing frequency selectivity will tend to produce the effects of a simple resonant filter; whereas both the shape and impulse response characteristics of the cochlear filters (at least as seen by cochlear nerve fibres) are characteristic of multipolar filters (see Evans 1989). Does Professor Ruggero have any idea how this higher order filtering is brought about? Is it possible that multipolar filtering could be brought about by means of interaction between the inherent (passive) basilar membrane filtering and the (active) positive feedback process?

M. A. RUGGERO. In response to Professor Evans' first question, I believe that upon uncoupling of the outer hair cell stereocilia from the tectorial membrane the basilar membrane response should certainly become 'passive' (and thus poorly frequency tuned) due to reduction or disappearance of receptor potentials, or electromotile response, or both. Indeed, such uncoupling may well be a common mechanism of cochlear damage during surgery preliminary to basilar membrane recordings. However, as I commented in the Discussion of the paper, the 'big unanswered question . . . is whether the mild intrinsic frequency tuning of the basilar membrane . . . is augmented by its interaction with the outer hair cells (and, indirectly, the tectorial membrane) or whether . . . outer hair cells are intrinsically sharply tuned'.

My response to the second question is I do not know how organ of Corti or basilar membrane filtering is brought about. The literature of cochlear modeling shows that simultaneously achieving both sensitivity and realistically wide frequency tuning is very difficult. The possibility that Professor Evans raises, namely of an interaction between passive and active processes, is not farfetched. Our data on the cochlear mechanical effects of furosemide (Ruggero & Rich 1991*b*) implies that the outer hair cells influence the frequency selectivity and sensitivity of basilar membrane vibration. If, as is likely, this influence is controlled by the receptor potentials of outer hair cells, then the process clearly involves a feedback loop. However, nobody knows whether the feedback is 'simple', whether it is positive or negative, or whether it includes AC or DC motors.

Sensory transduction and frequency selectivity in the basal turn of the guinea-pig cochlea

I. J. RUSSELL AND M. KÖSSL†

School of Biological Science, University of Sussex, Falmer, Brighton, Sussex BN1 9QG, U.K.

SUMMARY

Receptor potentials recorded from outer hair cells (OHC) and inner hair cells (IHC) in the basal high-frequency turn were compared. The DC component of the IHC receptor potential is maximized to ensure that IHCs can signal a voltage response to high-frequency tones. The OHC DC component is minimized so that OHCs transduce in the most sensitive region of their operating range. The phase and magnitude of OHC receptor potentials were recorded as an indicator of the magnitude and phase of the energy which is fed back to the basilar membrane to provide the basis for the sharp tuning and fine sensitivity of the cochlea to tones. IHC receptor potentials were recorded to assess the net effect of the feedback on the mechanics of the cochlea. It was concluded that OHCs generate feedback which enhances the IHC responses only at the best frequency. At frequencies below CF, IHC DC responses are elicited only when the OHC AC responses begin to saturate.

1. INTRODUCTION

The function of hearing in man is subserved in each cochlea by a single row of 3500 inner hair cells (IHCs) and 12 000 outer hair cells (OHCs) arranged in three rows. IHCs form synapses with the vast majority of afferent fibres in the auditory nerve whereas the OHCs have been described as the end organs of the efferent system (Davis 1983). The rows run parallel to the long axis of the spiral coil of the basilar membrane and the hair cells and their innervation are distributed tonotopically primarily as a consequence of the basilar membrane's mechanical tuning properties (see Evans, this symposium; Ruggero *et al.*, this symposium). Low-frequency hair cells are located at the apex of the cochlea and high-frequency hair cells are located in the basal turn. Hair cells are excited through displacements of the stereocilia bundle and the mechanosensitive channels are gated when the stereocilia are displaced towards the tallest row (Hudspeth & Corey 1977; Russell *et al.* 1986). The precise way in which the stereocilia of cochlear hair cells become displaced is not known but it is believed to occur as a consequence of shear displacements between the tectorial membrane and the basilar membrane which are the two principal structural components of the cochlear partition (Davis 1965). The rows of stereocilia of the OHCs in the mammalian cochlea are attached by their tips to the tectorial membrane and thus mechanically link the tectorial and basilar membranes. As a consequence of their strategic location in

the cochlear partition. OHCs play an essential role in the frequency tuning and sensitivity of the cochlea. After selective damage to the OHCs, the electrophysiological and mechanical responses of the cochlea to acoustic stimulation become insensitive, linear and broadly tuned (Liberman & Dodds 1984; Brown *et al.* 1989). This finding, together with the measurement of acoustic emissions from the cochlea (Kemp 1978) and the discovery that isolated OHCs are capable of rapid voltage-dependent motility (Brownell *et al.* 1985; Ashmore 1987), has led to the proposal that OHCs have an interactive role in sensory transduction in the cochlea (see Dallos (1988) for a review). More specifically, it has been suggested that OHCs feedback energy which overcomes viscous damping of the cochlear partion and provides the sharp frequency tuning of the cochlear responses (see, for example, Weiss (1982); Davis (1983); Neely & Kim (1983)).

It might be expected that the effectiveness of the proposed feedback depends on the mechanical properties of the basilar and tectorial membranes and on the gain and phase of the feedback process which has been associated with the OHC transducer. If, at low and moderate sound levels, OHCs operate in a true electromotor feedback loop then they will contribute to their own voltage responses and hence to the mechanical responses of the cochlear partition. For high levels of the acoustic stimulus the OHC transducer is saturated and the mechanical responses of the basilar membrane and the voltage responses of the OHCs are governed by the passive mechanical properties of the basilar membrane (Patuzzi *et al.* 1989; Zwislocki 1986). Thus, differences in the magnitude and phase of OHC voltage responses to tones at low and high levels should

† Present address: Zoological Institute, University of Munich, Luisenstrasse 14, 8000 Munich 2, F.R.G.

Phil. Trans. R. Soc. Lond. B (1992) **336**, 317–324
Printed in Great Britain

317

© 1992 The Royal Society and the authors

[23]

provide an indication of the magnitude and phase of the OHC feedback relative to the passive mechanical properties of the basilar membrane. A comparison between OHC and IHC responses from the same region of the cochlea should provide information about the relationship between OHC feedback and the net mechanical response of the cochlear partition which is reflected in the IHC DC receptor potentials (Russell & Sellick 1978). To this end, measurements were made from hair cells in the basal, high-frequency turn of the guinea-pig cochlea to tones within an octave of the CF of the hair cells (15–19 kHz) in an attempt to understand how electromechanical feedback contributes to the tuning of the cochlea. Before dealing with the question of frequency tuning in the cochlea, attention is drawn to the functional significance of the differences in the tone-evoked voltage responses of basal turn IHCS and OHCS.

2. IHC DC RESPONSES ARE MAXIMIZED

The resting potentials of IHCS recorded *in situ* are about -45 mV (Russell & Sellick 1978; Dallos 1986) and the relationship between the peak IHC voltage response to low-frequency tones and peak sound pressure (figure 1a, the transducer function) is a typical asymmetrical S-shaped curve which can be fitted by pairs of hyperbolic tangent functions (Russell & Sellick 1983; Dallos 1986; Russell & Kössl 1991) (figure 1b). The resting potential and the shape of the transducer function largely define the important functional characteristics of IHCS and the afferent fibres

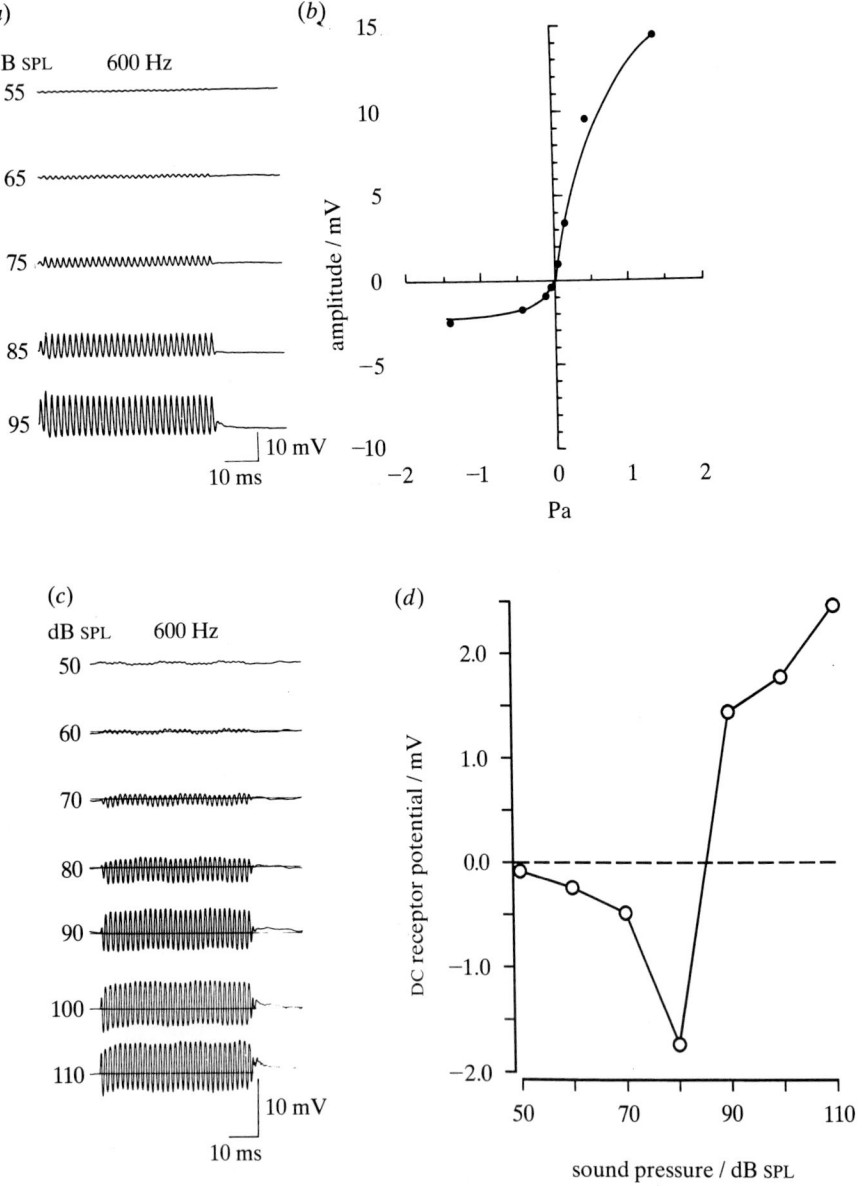

Figure 1. Intracellular receptor potentials recorded from (a) an IHC and (c) an OHC in the basal turn of the guinea-pig cochlea to 600 Hz tones at the sound pressures indicated by each trace. The peak positive and peak negative voltage responses of the IHC receptor potential as a function of the peak rarefaction and compression of the 600 Hz tone measured at the auditory meatus (transducer function) are shown in (b). The points are fitted by a pair of hyperbolic tangent functions. (d) The DC component of the OHC receptor potential as a function of the level of the 600 Hz tone. (a), (b) and (c) from Cody & Russell (1987).

with which they form synapses. At a resting potential of −45 mV, the voltage-dependent calcium conductance of the IHC basolateral membranes will be activated (Crawford & Kros 1990) and it is probable that calcium influx close to the presynaptic membrane of the IHC is instrumental in causing the spontaneous release of transmitter at the afferent synapses of the two spontaneously active classes of the three classes of afferent fibres which have so far been identified in the auditory nerve (Winter *et al.* 1990). The relatively depolarized IHC resting potential and spontaneous release of afferent transmitter ensure that the voltage responses of the IHCs are transmitted to the hind brain with maximum sensitivity through the modulation of a sustained level of transmitter and that a voltage threshold does not have to be exceeded before the afferent fibres are excited. The neural signalling of IHC voltage responses through the modulation of a sustained release of afferent synaptic transmitter provides the opportunity to signal negative as well as positive changes in sound pressure at levels close to the threshold of hearing for frequencies below about 1 kHz. Hence, at these levels afferent fibres can signal both rarefaction and compression in sound pressure (Palmer & Russell 1986). The asymmetrical transfer function provides the basis of the IHC DC receptor potential and the means to signal responses to high-frequency tones (Russell & Sellick 1978). The transducer function is shaped by the gating characteristics of the transducer channel which, for cochlear hair cells, can be modelled as a three state process with one

open and two closed states (Kros *et al.* 1991), and the voltage- and ion-dependent conductances of the basolateral channels (Kros & Crawford 1990; Dallos & Cheatham 1991). At rest, about 10% of the transducer conductance is open (Russell & Kössl 1991) and it is likely that the proportion of channels open at rest is governed by the concentration of intracellular free calcium (Crawford *et al.* 1991).

For frequencies above about 600 Hz, the AC receptor potential is attenuated by the low pass electrical characteristics of the hair cell so that the principal voltage responses of IHCs in the basal turn to tones at CF is the DC receptor potential (Russell & Sellick 1978; 1983). If the IHC afferent synapse is conventional, then transmitter release is primarily controlled by changes in presynaptic potential. This suggestion is supported by the close correspondence between phase-locking in the auditory nerve and the cut-off frequency of the IHC membrane time constant (Palmer & Russell 1986). For frequencies above the cut-off frequency, phase-locking declines according to the membrane time constant. However, for frequencies above about 3 kHz phase-locking is also limited by transmission at the afferent synapse (Weiss & Rose 1988), possibly due to limits which include those set by the speed of the calcium influx, transmitter mobilization and release, the kinetics of the post synaptic ligand gated channel and the spike generator.

3. OHC DC VOLTAGE RESPONSES ARE MINIMIZED

The voltage responses of OHCs in the basal turn of the guinea pig cochlea to tones close to their CF are remarkable for their small size (Russell *et al.* 1986; Cody & Russell 1987). At the threshold for neural excitation the magnitude of OHC voltage responses at CF are about 30 μV compared with 800 μV for IHCs (Russell & Sellick 1983; Sellick *et al.* 1983; Cody & Russell 1987; Russell & Kössl 1992*a, b*). The small size of the OHC voltage responses may be attributed to the almost symmetrical transducer function and the membrane time constant which attenuates voltage signals at frequencies above about 1 kHz (Dallos 1984; Cody & Russell 1987). The symmetry of the OHC transducer function at the CF can only by presumed from the absence of a DC voltage response to tones. At frequencies well below the cut-off of the membrane time constant, the OHC transducer function is much more symmetrical than the IHC transducer function but level dependent (figure 1*c, d*; Russell *et al.* 1986; Cody & Russell 1987). For low-level stimuli, OHCs generate predominantly hyperpolarizing responses, almost symmetrical responses to tones around 85 dB SPL and depolarizing responses at levels above this (figure 1*c, d*). OHCs in the apical turn of the guinea-pig cochlea exhibit similar level-dependent voltage responses to tones (Dallos *et al.* 1982) and very recently these level-dependent tone-evoked voltage responses have been compared with level-dependent length changes in isolated OHCs caused by transcellular current stimulation (Evans *et al.* 1988). The amplitude and polarity of the length change varies as

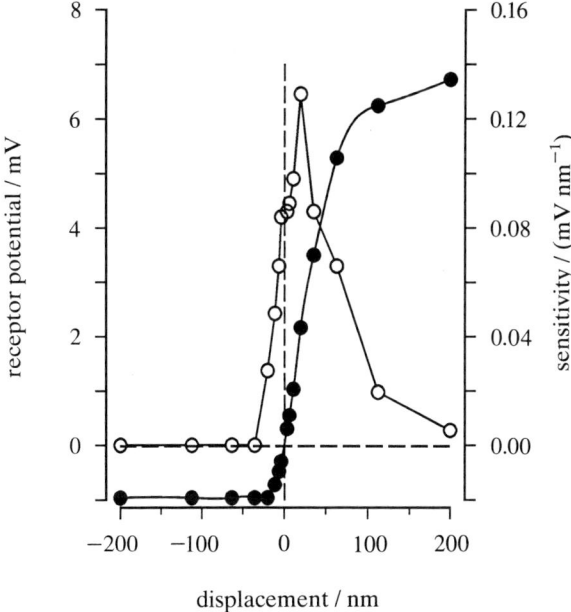

Figure 2. OHC receptor potential as a function of the displacement of the stereocilia bundle (solid circles). Sensitivity of OHC (open circles) based on the slope of the transducer function. Each point is the mean of measurements made from the receptor potentials recorded from five OHCs in an organ culture of a one-day post-natal mouse cochlea (the standard deviations do not exceed 12% of the mean). The dotted line represents the OHC resting membrane potential.

a function of current strength in a way analogous to the level dependency of tone-evoked voltage responses.

In neonatal mouse OHCs the steepest and hence most sensitive region of the transducer function is reached at about 20 nm displacement of the stereocilia in the excitatory direction (figure 2). This position corresponds to the operating point *in vivo* where the DC component of the receptor potential is at a minimum and the response is symmetrical. This feature of OHC transduction and the relatively hyperpolarized OHC resting membrane potential ensures maximum current flow through the transducer channels for a given displacement. Thus in contrast to the response characteristics of IHCs which are set to maximize the DC receptor potential and hence ensure a voltage response to a high-frequency tone, operating characteristics of OHCs ensure that OHCs operate in a most sensitive region of their operating range at the expense of not generating substantial voltage responses to tones at the CF and perhaps the opportunity to communicate these responses to the central nervous system.

The mechanism responsible for controlling the operating point of OHCs in the basal region of the

cochlea is not known. OHC receptor potentials (Russell *et al.* 1986) and receptor currents (Kros *et al.* 1991) in the organ culture of the mouse cochlea and in the apical turns of the guinea-pig cochlea (Dallos *et al.* 1982) are asymmetrical with a substantial DC component. It is possible that the hair bundle is biassed through interaction with the tectorial membrane so that at rest about 50% of the transducer conductance is open (Russell *et al.* 1986; Russell & Kössl 1991).

4. PHASE AND MAGNITUDE OF IHC AND OHC RECEPTOR POTENTIALS

The magnitude–level and phase–level functions of the AC voltage responses of OHCs (CF, 16 kHz) measured intracellularly and just extracellularly (figure 3) are remarkably similar. For frequencies below about one half an octave below the CF of an OHC, the magnitude of the AC component of the receptor potential increases linearly with levels below about 70 dB SPL. Above this level the slope of the AC–level function decreases and the receptor potential contains a DC component. At these frequencies and for low levels,

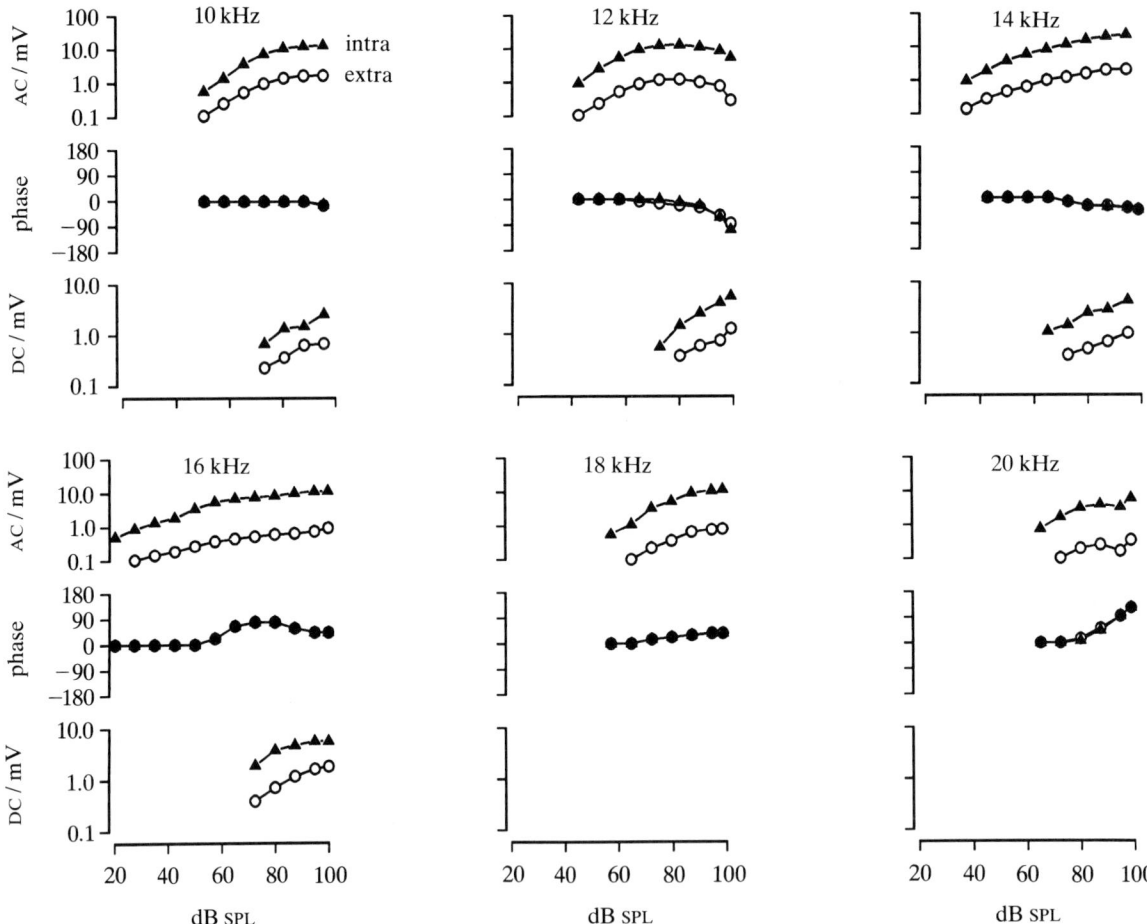

Figure 3. Tone-evoked voltage responses as functions of level (and at the frequencies shown) recorded intracellularly (solid triangle) and extracellularly (open circles) from an OHC. In each figure the magnitude of the AC response (upper), the phase of the AC response (middle) and the amplitude of the DC receptor potential (lower) is plotted as a function of tone level. Resting potential of OHC, −70 mV. Detection threshold at CF (16 kHz) was 20 dB SPL. The magnitudes of the responses are RMS values. Phase is measured with reference to the phase of the AC response at detection threshold (zero phase). Positive values of phase indicate leads and the data has been compensated for recording system and membrane time constants of 3.5 kHz and 1.1 kHz respectively. From Kössl & Russell (1992).

the phase of the AC component is almost independent of level. When the level exceeds about 70 dB SPL, the phase slightly, but measurably, lags the low-level response by a few degrees (e.g. up to $-45°$, see figure 3, 12 kHz) when the level of the tone exceeds about 70 dB SPL and the OHC AC component (and hence feedback to the cochlear partition) begins to saturate. For frequencies below the CF, it is only when the OHC AC responses begin to saturate that the neighbouring IHCS generate DC responses to the high-frequency tones (compare figures 3 and 4). For frequencies between one half of an octave below CF and CF, the phase of the OHC AC response suddenly jumps 180° and the magnitude of the AC response increases steeply in amplitude with increasing level when the level of the tone exceeds about 90 dB SPL (figure 2, 12 kHz; figure 4, 13 kHz). The phase reversal of the OHC AC response at very high levels of stimulation may indicate that a mechanical change has taken place in the cochlear partition. With respect to this, it has been proposed that the phase of excitation of OHCS is determined by the relationship between the rotational stiffness of the OHC stereocilia and the transverse stiffness of the tectorial membrane (Zwislocki 1986; 1988). A change in the relative values of these stiffnesses during stimulation with high-intensity tones could switch the phase of excitatory displacement of the OHC stereocilia from basilar membrane displacements towards scala tympani to displacements towards scala media (Mountain & Cody 1988).

At the CF, the detection threshold of the OHC AC response and IHC DC response and the initial slopes of the response–level functions are similar (figures 3 and 4). At sound levels around 60 dB SPL the AC signal begins to phase lead, amounting to approximately 90° at 70 dB SPL. Within the same range of levels, the OHC DC potentials first appear and the slope of the IHC DC response–level function becomes more shallow. The interpretation that has been put on these findings is that for low-level CF tones, the vibration of the basilar membrane in a sensitive cochlea phase-lags by 90° passive basilar membrane vibration without feedback from the OHCS. Thus IHCS are excited when the phase of the OHC AC response corresponds to basilar membrane velocity. With increasing levels of the tone, the relative contribution of the feedback from the OHCS is reduced with a corresponding progressive decrease in the phase lag of the basilar membrane motion. When the phase of the OHC feedback and presumably the effectiveness of the feedback in driving the basilar membrane is reduced, the slopes of the IHC DC response–level functions are also reduced. For frequencies below CF (e.g. figure 2, 12 kHz and 14 kHz; figure 4, 13 kHz) the phase of the low level AC response leads the high-level response which is opposite to the situation at the CF. At frequencies below the CF and at levels where the relative contribution of the OHC responses to basilar membrane vibration should be at a maximum, OHC feedback to the cochlear partition is not effective in exciting the IHCS. In fact it may even prevent excitation because the IHC DC response appears only when the OHC AC response begins to saturate and the phase of OHC excitation

approaches that of the passive basilar membrane. The observations presented above are in accordance with a model of frequency tuning in which OHCs contribute negative feedback to the cochlear partition which is reversed through a frequency-dependent phase delay (e.g. a low-pass filter) to become positive feedback at the CF (Mountain *et al.* 1983). That is electromotor feedback from the OHCS opposes basilar membrane displacement at frequencies away from the CF, but augments basilar membrane velocity at the CF.

5. THE EFFECT OF LOW-FREQUENCY TONES ON HAIR CELL HIGH-FREQUENCY RESPONSES

It has been proposed that by minimizing the DC component of the receptor potential, OHCs optimize electromechanical feedback by keeping the operating point in the steepest part of the transducer function (Russell *et al.* 1986; Russell & Kössl 1991). Any disturbance of the OHC transducer function away from this operating point should result in a decrease in sensitivity of cochlear responses, particularly at frequencies close to the CF. Decreases in frequency tuning, particularly at the CF have indeed been measured in the responses of hair cells, as a result of the modulation of the responses to high-frequency tones by low-frequency tones (Patuzzi & Sellick 1984).

OHC AC responses to tones at frequencies below CF are suppressed by simultaneously presented 80 dB SPL 100 Hz tones. Suppression is not associated with a phase change in the AC response and suppression of the high-frequency response disappears when the level of the high-frequency tone exceeds about 70 dB SPL and the AC response begins to saturate (figure 4). At these frequencies, IHC DC receptor potentials are not suppressed by the 100 Hz tone because they do not appear until the OHC AC response has begun to saturate and is no longer inhibited by the 100 Hz tone. At the CF, OHC responses are suppressed by the 100 Hz tone at levels below about 70 dB SPL. Suppression of the OHC AC response at the CF is associated with a phase lead of about 90° (figure 4) and by suppression of the DC response in neighbouring IHCS. It is suggested that these observations support the idea that OHCs feedback energy to the cochlea partition at the CF at a phase which would correspond to the velocity of the basilar membrane without feedback. Suppression of the OHC AC response at CF at levels below about 70 dB SPL causes a change in the phase of excitation of the OHC AC response so that it now corresponds to the displacement of the basilar membrane without feedback and is associated with suppression of the IHC DC response.

6. HAIR CELL ISORESPONSE TUNING CURVES

It was suggested above that, at low sound levels, the AC response of the OHC receptor potential provides a measure of the mechanical feedback to the cochlear partition and the DC response of the IHC receptor

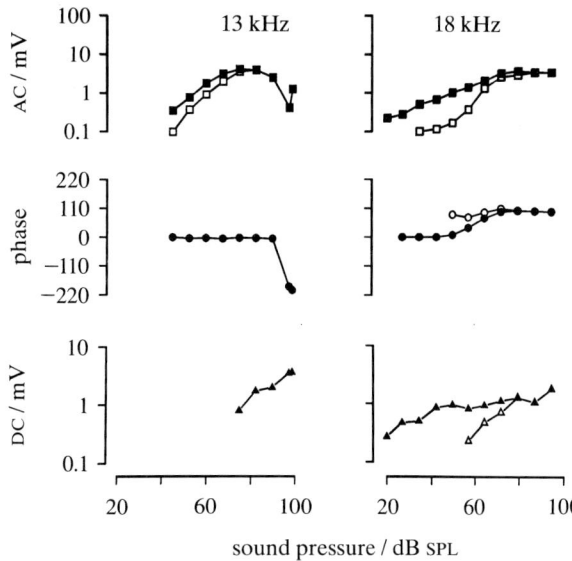

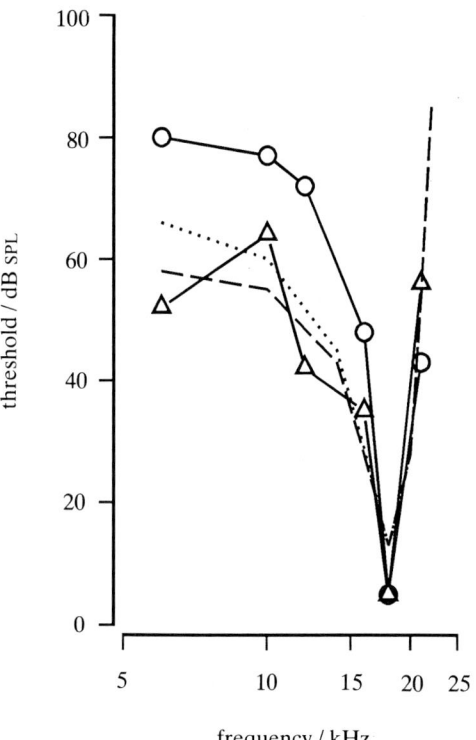

Figure 4. Magnitude, phase and DC response–level functions based on extracellular recordings close to an IHC at the frequencies indicated. In each figure the magnitude of the AC response (upper), the phase of the AC response (middle) and the DC receptor potential (lower) is plotted as a function of tone level. Solid symbols represent the response to the high frequency tone alone, open symbols represent the response to the combined 100 Hz, 80 dB SPL tone and high-frequency tone. The magnitude of the AC response is corrected by 6 dB per octave from 4 kHz to compensate for the low-pass filtering due to the recording system. From Russell & Kössl (1992a).

Figure 5. Isoresponse tuning curves measured intracellularly from an OHC and an IHC in the 18 kHz region of the same cochlea and isoresponse measurements of the displacement and velocity of the 18 kHz region of the guinea-pig basilar membrane taken from Sellick *et al.* (1982). The criteria for the isoresponse tuning curves were 0.5 mV (OHC AC, triangles), 0.8 mV (IHC DC, circles), 0.044 mm s^{-1} (basilar membrane velocity, dotted line), 3.5 Å (basilar membrane displacement, dashed line). The resting potentials of the IHC and OHC were -45 mV and -65 mV respectively. Threshold loss after exposure of the basilar membrane was less than 5 dB SPL. The AC responses were compensated for the recording system time constant (3 kHz). CF, 15 kHz; solid circles, IHC DC intracellularly; open triangles, OHC AC extracellularly.

potential is an indicator of the net effect of this feedback on the mechanics of the cochlear partition. Furthermore it has been proposed that OHC feedback reverses from negative feedback at frequencies below the CF to positive feedback at the CF (Mountain *et al.* 1983). Thus, it might be expected that the isoresponse frequency tuning curves of IHCs and OHC may differ from each other and from isoresponse tuning curves of the basilar membrane mechanics. On the basis of intracellular recordings from adjacent IHCs and OHCs without a measurable change in cochlear sensitivity between and following the recordings, the characteristics of IHC and OHC isoresponse tuning curves are very similar in the tip region (Russell *et al.* 1986; Cody & Russell 1987; Kössl & Russell 1992) (figure 5). The bandwidth of the tuning curve, measured 10 dB from the tip ($Q_{10\,\mathrm{dB}}$) and high and low frequency slopes are similar to those which have been measured in primary afferent fibres and the basilar membrane in the basal turn of the guinea-pig cochlea (Sellick *et al.* 1982, 1983; Nuttall *et al.* 1991). In all respects, the tuning curves of IHCs and nerve fibres are almost identical which may indicate that synaptic transmission across the IHC afferent synapse is frequency independent, at least for frequencies above a few kHz. However, IHC and OHC tuning curves are dissimilar in that the difference between the tip and the low frequency 'tail' of OHC tuning curves is about 15 dB less than that of IHC and neural tuning curves. This property reflects the finding that, for frequencies on the low-frequency

tail of the tuning curve, IHC DC receptor potentials are not generated until the OHC AC responses begin to saturate. In this respect OHC tuning curves resemble isodisplacement tuning curves of the basilar membrane (figure 5) where it has been observed that the sharp transition between the tip and the low frequency tail, which characterizes IHC and neural tuning curves, is absent (Sellick *et al.* 1983). It remains to be seen if the introduction of more sensitive laser Doppler velocimetry and optical techniques will result in basilar membrane isoamplitude tuning curves which more closely resemble neural tuning curves. The limited data which is currently available does not resolve this issue (Nuttall *et al.* 1991). It should be pointed out that differences exist between laboratories and species in the measurement of basilar membrane mechanics which have been discussed elsewhere (Robles *et al.* 1986; Ruggero *et al.* 1986; Ruggero & Rich 1991). For example, there is good agreement between isodisplacement basilar membrane tuning curves and neural tuning curves in the 7–10 kHz region of the chinchilla cochlea (Robles *et al.* 1986).

On the basis of the available evidence to date it is suggested that in the guinea-pig cochlea the OHC tuning curves closely reflect iso-displacement tuning curves of the basilar membrane and that IHC and neural tuning curve reflect the net radial shear displacement between the basilar membrane and tectorial membrane.

7. CONCLUSION

There is increasing evidence from a wide variety of sources to show that cochlear sensitivity and tuning both depend on feedback from OHCs. It is clear that the phase of feedback relative to the passive motion of the basilar membrane is of crucial importance for the feedback to be effective. What is not clear is how the phase of the feedback is determined or indeed what is the source of the feedback. At present there are two candidates for the feedback process. One of these is the fast electromotile process which some argue may not be fast enough or large enough at high frequencies to provide the feedback (Santos-Sacchi 1989, 1990; Hudspeth 1990; Dallos 1991). The other is a proposed voltage or displacement dependent stiffness change in the hair bundle itself (see Dallos 1991). Time, as they say, will tell.

This research was supported by grants from the MRC, Wolfson Foundation and Hearing Research Trust.

REFERENCES

Ashmore, J.F. 1987 A fast motile event in outer hair cells isolated from the guinea pig cochlea. *J. Physiol., Lond.* **385**, 207–242.

Brown, A.M., McDowell, B. & Forge, A. 1989 Acoustic distortion products can be used to monitor the effects of chronic gentamicin treatment. *Hear. Res.* **42**, 143–156.

Brownell, W.E., Bader, C.R., Bertrand, D. & de Ribaupierre, Y. 1985 Evoked mechanical responses of isolated cochlear outer hair cells. *Science, Wash.* **227**, 194–196.

Cody, A.R. & Russell, I.J. 1987 The responses of hair cells in the basal turn of the guinea-pig cochlea to tones. *J. Physiol., Lond.* **383**, 551–569.

Crawford, A.C. & Kros, C.J. 1990 A fast calcium current with a rapidly inactivating component in isolated inner hair cells of the guinea pig. *J. Physiol., Lond.* **420**, 90P.

Crawford, A.C., Evans, M. & Fettiplace, R. 1991 The actions of calcium on the mechano-electric transducer current of turtle hair cells. *J. Physiol., Lond.* **434**, 369–398.

Dallos, P. 1984 Some electrical circuit properties of the organ of Corti. II. Analysis including reactive elements. *Hear. Res.* **14**, 281–291.

Dallos, P. 1986 Neurobiology of cochlear inner and outer hair cells, Intracellular recordings. *Hear. Res.* **22**, 185–198.

Dallos, P. 1988 Cochlear neurobiology, some key experiments and concepts of the past two decades. In *Auditory function, neurobiological basis of hearing* (ed. G. M. Edelman, W. E. Gall & W. M. Cowan), pp. 153–188. New York: Wiley.

Dallos, P. 1992 Neurobiology of cochlear hair cells. In *Auditory physiology and perception* (ed. Y. Cazels, L. Demany & K. Horner). Oxford: Pergamon Press. (In the press.)

Dallos, P. & Cheatham, M.A. 1991 Effects of electrical polarization on inner hair cell receptor potentials. *J. acoust. Soc. Am.* **87**, 1636–47.

Dallos, P., Santos-Sacchi, J. & Flock, Å. 1982 Intracellular recordings from cochlear outer hair cells. *Science, Wash.* **218**, 582–584.

Davis, H. 1965 A model for transducer action in the cochlea. *Cold Spring Harb. Symp. quant. Biol.* **30**, 181–190.

Davis, H. 1983 An active process in cochlear mechanics. *Hear. Res.* **9**, 79–90.

Evans, R.N., Hallworth, R. & Dallos, P. 1988 Asymmetries in motile responses of outer hair cells in stimulated in vivo conditions. In *Cochlear mechanisms, structure function and models* (ed. J. P. Wilson & D. T. Kemp), pp. 205–206. London: Plenum Press.

Hudspeth, A.J. 1990 How the ear's works work. *Nature, Lond.* **341**, 397–404.

Hudspeth, A.J. & Corey, D.P. 1977 Sensitivity, polarity and conductance change in the response of vertebrate hair cells to controlled mechanical stimuli. *Proc. natn. Acad. Sci. U.S.A.* **74**, 2407–2441.

Kemp, D.T. 1978 Stimulated acoustic emissions from within the human auditory system. *J. acoust. Soc. Am.* **64**, 1386–1391.

Kössl, M. & Russell, I.J. 1992 The phase and magnitude of hair cell receptor potentials and frequency tuning in the guinea pig cochlea. *J. Neurosci.* (In the press.)

Kros, C.J. & Crawford, A.C. 1990 Potassium currents in inner hair cells isolated from the guinea pig cochlea. *J. Physiol., Lond.* **421**, 263–291.

Kros, C.J., Rüsch, A., Richardson, G.P. & Russell, I.J. 1991 Transducer currents in outer hair cells in cultures of neonatal mice. *J. Physiol., Lond.* **446**, 112P.

Liberman, M.C. & Dodds, L.W. 1984 Single-neuron labeling and chronic cochlear pathology. III. Stereocilia damage and alterations of threshold tuning curves. *Hear. Res.* **16**, 55–74.

Mountain, D.C., Hubbard, A.E. & McMullen, T.A. 1983 Electromechanical processes in the cochlea. In *Mechanics of hearing* (ed. E. de Boer & M. A. Viergever), pp. 119–126. Delft University Press.

Mountain, D.C. & Cody, A.R. 1989 Mechanical coupling between inner and outer hair cells in the mammalian cochlea. In *Cochlear mechanisms, structure function and models* (ed. J. P. Wilson & D. T. Kemp), pp. 153–160. London: Plenum Press.

Neely, S.T. & Kim, D.O. 1983 An active cochlear model showing sharp tuning and high sensitivity. *Hear. Res.* **9**, 123–130.

Nuttall, A.L., Dolan, D.F. & Avinash, G. 1991 Laser Doppler velocimetry of the basilar membrane. *Hear. Res.* **51**, 203–214.

Palmer, A.R. & Russell, I.J. 1986 Phase-locking in the auditory nerve of the guinea-pig and its relation to the receptor potential of inner hair-cells. *Hear. Res.* **24**, 1–15.

Patuzzi, R. & Sellick, P.M. 1984 The modulation of the sensitivity of the mammalian cochlea by low-frequency tones. II. Inner hair cell receptor potentials. *Hear. Res.* **13**, 9–18.

Patuzzi, R., Yates, G.K. & Johnstone, B.M. 1989 Outer hair cell receptor currents and sensorineural hearing loss. *Hear. Res.* **42**, 47–72.

Robles, L., Ruggero, M.A. & Rich, N.C. 1986 Basilar membrane mechanics at the base of the chinchilla cochlea. I. Input-output functions, tuning curves, and response phases. *J. acoust. Soc. Am.* **80**, 1364–1374.

Ruggero, M.A. & Rich, N.C. 1991 Application of a commercially-manufactured Doppler-shift laser velocimeter to the measurement of basilar membrane vibration. *Hear. Res.* **51**, 215–230.

Ruggero, M.A., Robles, L. & Rich, N.C. 1986 Basilar membrane mechanics at the base of the chinchilla cochlea. II. Responses to low-frequency tones and the relationship to microphonics and spike initiation in the VIII nerve. *J. acoust. Soc. Am.* **80**, 1375–1383.

Russell, I.J. & Kössl, M. 1991 The voltage responses of hair cells in the basal turn of the guinea-pig cochlea. *J. Physiol., Lond.* **435**, 493–511.

Russell, I.J. & Kössl, M. 1992*a* Modulation of hair cell responses to tones by low frequency biassing of the basilar membrane in the guinea pig cochlea. *J. Neurosci.* (In the press.)

Russell, I.J. & Kössl, M. 1992*b* Voltage responses to tones of outer hair cells in the basal turn of the guinea-pig cochlea: significance for electromotility and desensitization. *Proc. R. Soc. Lond.* B, **247**, 97–105.

Russell, I.J. & Sellick, P.M. 1978 Intracellular studies of hair cells in the mammalian cochlea. *J. Physiol., Lond.* **284**, 261–290.

Russell, I.J. & Sellick, P.M. 1983 Low-frequency characteristics of intracellularly recorded receptor potentials in guinea-pig cochlear hair cells. *J. Physiol., Lond.* **338**, 179–206.

Russell, I.J., Cody, A.R. & Richardson, G.P. 1986 The responses of inner and outer hair cells in the basal turn of the guinea-pig cochlea and in the mouse cochlea grown in vitro. *Hear. Res.* **22**, 199–216.

Santos-Sacchi. J. 1989 Asymmetry in voltage dependent movements of isolated outer hair cells from the organ of Corti. *J. Neurosci.* **9**, 2954–2952.

Santos-Sacchi, J. 1990 Fast outer hair cell motility, how fast is fast? In *Mechanics and biophysics of hearing* (ed. P. Dallos, C. D. Geisler, J. W. Mathews, M. Ruggero & C. R. Steele), pp. 69–75. New York: Springer.

Sellick, P.M., Patuzzi, R. & Johnstone, B.M. 1982 Measurement of basilar membrane motion in the guinea pig using the Mössbauer technique. *J. acoust. Soc. Am.* **61**, 133–149.

Sellick, P.M., Patuzzi, R. & Johnstone, B.M. 1983 Comparison between the tuning properties of inner hair cells and basilar membrane motion. *Hear. Res.* **10**, 93–100.

Winter, I.M., Robertson, D. & Yates, G.K. 1990 Diversity of characteristic frequency rate-intensity functions in guinea-pig auditory nerve fibres. *Hear. Res.* **45**, 203–220.

Weiss, T.F. 1982 Bidirectional transduction in vertebrate hair cells, A mechanism for coupling mechanical and electrical processes. *Hear. Res.* **7**, 353–360.

Weiss, T.F. & Rose, C. 1988 A comparison of synchronization filters in different auditory receptor organs. *Hear. Res.* **33**, 175–180.

Zwislocki, J.J. 1986 Analysis of cochlear mechanics. *Hear. Res.* **22**, 155–169.

Zwislocki, J.J. 1988 Phase reversal of OHC response at high sound intensities. In *Cochlear mechanisms, structure function and models* (ed. J. P. Wilson & D. T. Kemp), pp. 163–168. London: Plenum Press.

Evidence that adaptation of suppression cannot account for auditory enhancement or enhanced forward masking

BEVERLY A. WRIGHT[1] AND DENNIS McFADDEN[2]

[1] Psychoacoustics Laboratory, Department of Psychology, University of Florida, Gainesville, Florida 32611, U.S.A.
[2] Department of Psychology, University of Texas, Austin, Texas 78712, U.S.A.

SUMMARY

Delaying the onset of a signal relative to the onset of a simultaneous notched masker often improves the ability of listeners to 'hear out' the signal at both threshold and suprathreshold levels. Viemeister & Bacon (J. acoust. Soc. Am., **71**, 1502–1507 (1982)) suggested that such auditory enhancement effects could be accounted for if the suppression produced by the masker on the signal frequency adapted, thereby releasing the signal from suppression. In support of their hypothesis, Viemeister & Bacon reported that a masker preceded by an enhancer having no component at the signal frequency produced more forward masking than did the masker by itself. Here evidence is provided from five new experiments showing that adaptation of psychophysical two-tone suppression is inadequate to account either for auditory enhancement effects or for the enhanced forward masking demonstrated by Viemeister & Bacon.

1. INTRODUCTION

When a group of harmonics of equal amplitude are gated on and off together, it is difficult to 'hear out' a single target harmonic. However, if the onset of the target is delayed relative to the onset of the other harmonics, the detectability of the target improves with increasing onset delay (e.g. Viemeister 1980). Similar results are obtained with non-harmonic stimuli (see Viemeister, 1980). Such improvements in both the suprathreshold and threshold salience of a delayed target have been referred to as auditory enhancement (see, for example, Viemeister 1980; Summerfield *et al.* 1987), but here the term signal enhancement will be used to emphasize that the signal is enhanced in simultaneous masking experiments.

Most proposed explanations for signal enhancement have appealed to physiological short-term adaptation (e.g. Smith and Zwislocki 1975), but Viemeister & Bacon (1982) suggested an intriguing alternative. They hypothesized that signal enhancement was the result, not of the reduced masking ability of adapted, non-target maskers, but rather of a true gain in the effective level of the delayed target or signal, caused by adaptation of the psychophysical suppressive effect of the masker on that signal. According to this hypothesis, the psychophysical level of the masker remains unchanged throughout its timecourse, but the suppressive power of the masker declines over time. Hence, a delayed signal is more detectable because it is introduced into a partially, or completely, unsuppressed region. Viemeister & Bacon acknowledged that this proposed adaptation of psychophysical sup-

pression could not be mediated at the level of the 8th nerve, where it has been shown that physiological two-tone suppression does not adapt (Liff & Goldstein 1970).

In apparent support of their hypothesis of adaptation of suppression, Viemeister & Bacon (1982) reported that a harmonic complex produced approximately 8 dB more forward masking when it was preceded by an enhancer that was a longer version of the same complex, but with the component at the signal frequency deleted. Because the enhancer in Viemeister & Bacon's task increased the effectiveness of their forward masker, the difference in forward-masked thresholds obtained with the unenhanced and enhanced maskers will be referred to as masker enhancement.

Here the results of five experiments are reported that confirm and extend Viemeister & Bacon's report of masker enhancement. However, the present data appear to indicate (i) that signal and masker enhancement may not be as closely related as has previously been assumed, and (ii) that neither signal nor masker enhancement can be accounted for by adaptation of psychophysical two-tone suppression.

2. GENERAL METHOD

The same six normal-hearing subjects served in all five experiments. All had a minimum of one year's experience on other psychoacoustic tasks, and all of their data were monitored for evidence of learning. The procedure was adaptive, two-interval, forced-choice, with feedback. Thresholds are expressed as the

Phil. Trans. R. Soc. Lond. B (1992) **336**, 325–329
Printed in Great Britain

325

signal level estimated to be required for 79% correct detections. The threshold for each subject in each condition was based on the average of 4–18 blocks of 60 trials each. Standard errors of the mean were typically less than 1 dB within subjects. All listening was monotic. All waveforms were gated with a cosine-squared rise-decay time of 10 ms.

3. EXPERIMENTS AND RESULTS

(a) Psychophysical two-tone suppression

Viemeister & Bacon (1982) suggested that adaptation of suppression was the mechanism responsible both for signal enhancement in simultaneous masking and masker enhancement in forward masking. Therefore, in the first experiment, psychophysical two-tone suppression (see, for example, Houtgast 1972; Shannon 1976) was measured in the six subjects. The task was forward masking. The signal was a 1000 Hz tone, 20 ms in duration. The masker was a 1000 Hz tone having a level of 50 dB SPL, and the suppressor was a 1150 Hz tone having a level of 70 dB SPL. The common duration of the masker and suppressor was either 50, 100, or 500 ms. The onset of the signal coincided with the offset of the masker and suppressor. A contralateral wideband noise (300–4700 Hz, 40 dB SPL overall) was gated on and off with the masker or suppressor to help reduce any confusion effects that might have contributed to the results (see Moore & Glasberg 1982; Neff 1986). The amount of suppression was calculated as the difference in signal threshold obtained when the masker was presented alone and when the masker and suppressor were presented together.

The six subjects fell into two groups of three each based both upon their signal thresholds in the various conditions, and the magnitudes of their two-tone suppression effects. Three subjects consistently showed the lowest thresholds in the masker-only, suppressor-only, and masker-plus-suppressor conditions. These same subjects also showed typical amounts of psychophysical two-tone suppression, averaging 4, 5 and 8 dB for the 50, 100 and 500 ms stimuli, respectively. The other three subjects always had the highest thresholds in the various conditions, and showed small or even negative two-tone suppression. For these subjects, the average amount psychophysical suppression was about −6, −6 and −1 dB for the 50, 100 and 500 ms stimuli, respectively. (Note that for both groups, the amount of psychophysical suppression increased, or became less negative, with increasing stimulus duration. We return to this point in a later section.)

A close relationship between psychophysical suppression, signal enhancement, and masker enhancement is implicit in Viemeister & Bacon's (1982) proposal of adaptation of suppression. It was therefore predicted that the three subjects who had normal suppression would show the largest amounts of signal and masker enhancement, and that the three subjects who had negative suppression would show the smallest amounts. The results of the following experiments did not meet these predictions.

(b) Signal enhancement

In the second experiment, signal enhancement was measured in the same six subjects who had participated in the experiment on two-tone suppression. The task was simultaneous masking. Prompted by the stimuli used in the masker-enhancement experiments described below, in this signal-enhancement experiment, the masker covered the frequency range from 100–10 000 Hz and contained a spectral notch, 800 Hz in width, centred at 1000 Hz. The signal was a narrowband noise, 800 Hz in width, arithmetically centred at 1000 Hz. The duration of the signal was always 62 ms. The signal and masker were gated off together, but the onset of the masker preceded the onset of the signal by a variety of fringe durations. The spectrum level of the masker was 33 dB SPL. The results are shown separately for the six subjects in figure 1.

For all subjects, the signal became easier to hear as its onset was delayed from the onset of the masker. However, the six subjects fell into the same two distinct groups in terms of their signal thresholds and the magnitudes of their improvements with increasing signal delay. The three subjects who had normal two-tone suppression (open symbols) again showed the lowest thresholds, and also showed the smallest amount of signal enhancement, averaging about 7 dB. The three subjects who had negative two-tone suppression (filled symbols) again showed the highest thresholds, and also showed the largest amount of signal enhancement, averaging about 24 dB. These results are in sharp contrast to the predicted relationship between psychophysical two-tone suppression and signal enhancement, based upon Viemeister & Bacon's (1982) hypothesis of adaptation of suppression.

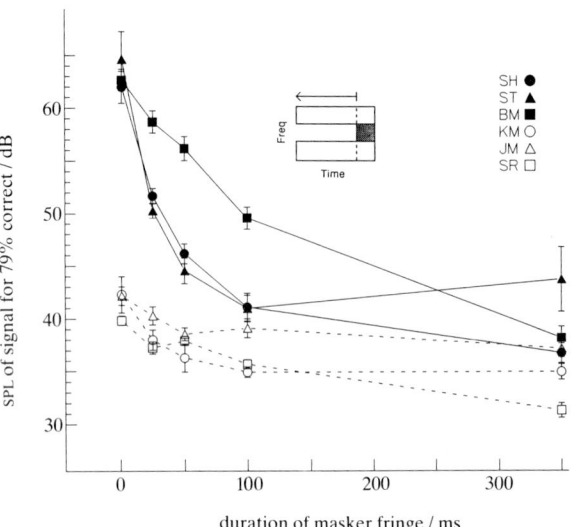

Figure 1. Signal enhancement in simultaneous masking for the three subjects who showed normal psychophysical two-tone suppression (open symbols) and the three subjects who showed negative psychophysical two-tone suppression (filled symbols).

(c) *Masker enhancement*

In the final three experiments, masker enhancement was measured in the same six subjects. The task was forward masking. The signal was a 1000 Hz tone, 20 ms in duration. The masker was a wideband noise covering the frequency range from 100 to 10 000 Hz, and the enhancer that preceded the masker covered the same frequency range as the masker, but contained a spectral notch of various widths centered on the signal frequency. The typical duration of the masker was 62 ms, and that of the enhancer was 350 ms. Both the masker and the enhancer were presented at a spectrum level of 33 dB SPL. The results of all three experiments were at odds with the hypothesis of adaptation of suppression.

(i) *Notch width in the enhancer*

In the first experiment on masker enhancement, the relative effectiveness of enhancers having a variety of notch widths was examined. Figure 2a shows the result for the three subjects with normal two-tone suppression and figure 2b shows the results for the three subjects with negative two-tone suppression. All six subjects performed similarly. For both groups, thresholds were higher when the masker was enhanced (squares) than when the masker was presented alone (dotted line), confirming the central

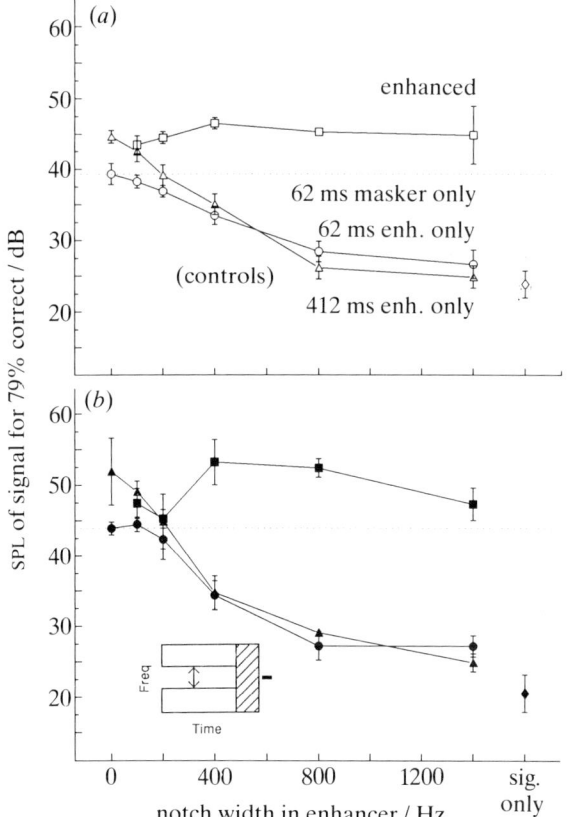

result of Viemeister & Bacon (1982). The amount of masker enhancement was greatest for both groups (averaging about 7 dB) when the notch width in the enhancer was 400 Hz, but there was considerable masker enhancement for wider notch widths. The difference in the total duration of the stimuli across the masker-only and enhanced conditions appears to have contributed substantially to the observed masker enhancement only for the narrowest notch widths. This is because thresholds with the wider notch widths were fairly similar when only the enhancer was presented for the duration of the masker-only (62 ms, circles) or for the duration of the enhanced (412 ms, triangles) conditions. As in the experiments above, here, and in the following two experiments, the subjects who showed normal two-tone suppression had somewhat lower thresholds than the subjects who showed negative two-tone suppression. However, because this difference was roughly constant across conditions, the amount of masker enhancement was nearly always within 2 dB across the two groups.

The results of this experiment appear to contradict the proposal that adaptation of two-tone suppression can account for masker enhancement, and therefore signal enhancement. Psychophysical two-tone suppression is greatest when the frequency of the suppressor is about 1.15 times the frequency of the masker and signal (Shannon 1976), and declines as that separation is increased. To the extent that the present results are influenced by two-tone suppression, it would therefore be expected that masker enhancement should be greatest with the notch width of 200 Hz (where the upper edge of the notch was closest to the ideal ratio of 1.15), and should decrease thereafter. However, masker enhancement was minimal with the 200-Hz notch, but was substantial at the wider notch widths.

(ii) *Enhancer duration*

In the second experiment on masker enhancement, the growth of the effect of the enhancer upon the masker was investigated by varying the duration of the enhancer. The duration of the masker was always 62 ms. The notch width in the enhancer was 800 Hz. The results were again similar across the two groups of subjects. For both groups, thresholds were higher in the enhanced than in the masker-only condition for all durations of the enhancer, and the difference in the threshold between the two conditions increased from approximately 3 to 9 dB as the enhancer duration was increased from 25 to 450 ms. These results also appear to be inconsistent with the proposal that adaptation of suppression could account for both masker and signal enhancement, because the amount of masker enhancement increased over the same timecourse that psychophysical two-tone suppression increases (see Weber & Green 1978; suppression experiment above).

(iii) *Masker duration*

In the final experiment on masker enhancement, the persistence of the enhancer's effect on the wideband masker was examined by varying the duration of the masker. The duration of the enhancer was always

Figure 2. Masker enhancement in forward masking as a function of the notch width in the enhancer for (a) the three subjects who showed normal psychophysical two-tone suppression (open symbols) and (b) the three subjects who showed negative psychophysical two-tone suppression (filled symbols).

[33]

350 ms, and the notch width in the enhancer was 800 Hz. Again the results were similar across the two groups of subjects. For both groups, thresholds were consistently higher in the enhanced than in the masker-only conditions. In the masker-only condition, thresholds increased by about 8.5 dB as the duration of the masker was increased from 22 to 302 ms. In the enhanced condition, however, thresholds increased by approximately 7 dB as the masker duration was increased from 22 to 102 ms, but declined by about 3 dB as the masker duration was increased from 102 to 302 ms. Therefore, the amount of masker enhancement was greatest for masker durations of 42 to 102 ms, reaching a peak of about 7 dB, and then decreased to about 3 dB for the 302 ms masker duration.

These results also do not support Viemeister & Bacon's (1982) proposal that adaptation of suppression can account for masker and signal enhancement. Psychophysical suppressors lose their ability to suppress when they are separated from the signal by only 50 ms (Weber & Green 1978), but enhancers have the ability to produce masker enhancement when they are separated from the signal by as much as 300 ms.

4. CONCLUSIONS

In summary, the results of five experiments appear to indicate that adaptation of psychophysical two-tone suppression cannot account for either signal or masker enhancement, and that signal and masker enhancement may not be as closely related as has previously been supposed. Perhaps the strongest evidence for the dissociation between adaptation of psychophysical two-tone suppression and signal enhancement was that the three subjects who showed negative two-tone suppression exhibited the largest amount of signal enhancement. The close relationship between signal and masker enhancement was also brought into question because subjects showing markedly different amounts of signal enhancement showed very similar amounts of masker enhancement. Finally, the parametric manipulations of the notch width in the enhancer, the duration of the enhancer, and the duration of the masker in the masker-enhancement experiments all were inconsistent with previous results obtained with psychophysical two-tone suppression, and thus also appeared to indicate that neither signal nor masker enhancement could be accounted for by an adaptation of suppression.

We thank C. A. Champlin and E. G. Pasanen for their helpful discussions. E. G. Pasanen provided technical assistance.

REFERENCES

Houtgast, T. 1972 Psychophysical evidence for lateral inhibition in hearing. *J. acoust. Soc. Am.* **51**, 1885–1894.

Liff, H. & Goldstein, M.J. Jr 1970 Peripheral inhibition in auditory fibers in the frog. *J. acoust. Soc. Am.* **47**, 1538–1547.

Moore, B.C.J. & Glasberg, B.R. 1982 Contralateral and ipsilateral cueing in forward masking. *J. acoust. Soc. Am.* **71**, 942–945.

Neff, D. 1986 Confusion effects with sinusoidal and narrow-band noise forward maskers. *J. acoust. Soc. Am.* **79**, 1519–1529.

Shannon, R.V. 1976 Two-tone unmasking and suppression in a forward-masking situation. *J. acoust. Soc. Am.* **59**, 1460–1470.

Smith, R.L. & Zwislocki, J.J. 1975 Short-term adaptation and incremental responses in single auditory-nerve fibers. *Biol. Cybern.* **17**, 169–182.

Summerfield, Q., Sidwell, A. & Nelson, T. 1987 Auditory enhancement of changes in spectral amplitude. *J. acoust. Soc. Am.* **81**, 700–708.

Viemeister, N.F. 1980 Adaptation of masking. In *Psychophysical, physiological and behavioral studies in hearing* (ed. G. van den Brink & F. A. Bilsen), pp. 190–197. Delft University Press.

Viemeister, N.F. & Bacon, S.P. 1982 Forward masking by enhanced components in harmonic complexes. *J. acoust. Soc. Am.* **71**, 1502–1507.

Weber, D.L. & Green, D.M. 1978 Temporal factors and suppression effects in backward and forward masking. *J. acoust. Soc. Am.* **64**, 1392–1399.

Discussion

B. C. J. MOORE (*Department of Experimental Psychology, University of Cambridge, U.K.*). In the interpretation of her results, I think that Dr Wright should carefully consider the possible cues used by the subjects in the various tasks. Whenever I see large individual differences between normally hearing subjects, I suspect that these arise from differences in the use of cues or processing strategies, rather than from differences in peripheral auditory processing.

In the case of 'unmasking' effects in forward masking, we have presented evidence that these can be strongly affected by the cues available to the subjects. In the 'reference' condition (on-frequency narrowband masker alone) thresholds may be high because the subjects lack an effective cue to distinguish the signal from the masker. Adding extra components to the masker can provide such a cue, thereby reducing thresholds (Moore 1980a, b) 1981; Terry & Moore 1977). Reductions in threshold in such cases should not be taken as indicating physiological suppression. Subjects may fail to show an unmasking effect for two reasons; firstly, they may 'find' an adequate cue in the reference condition, so that the added masker component provides no extra cue; secondly, they may not make effective use of the potential cue provided by the added masker component. The addition of a contralateral cue may be sufficient to provide an effective cue for some subjects (Moore & Glasberg 1982) but not for all subjects.

In the experiments showing 'enhanced' forward masking, it is possible that part of the enhancement effect arises from a kind of 'distraction'. Close to threshold, the perceptual cue associated with the presence of the signal is rather subtle. When the spectrum of the masker is changed substantially just before the signal occurs, the highly salient perceptual change associated with this may distract the subject from the subtle change associated with the signal, thereby raising threshold. A related explanation has been proposed by Bacon & Moore (1987) to account for certain temporal effects in simultaneous masking.

References

Bacon, S.P. & Moore, B.C.J. 1987 "Transient masking" and the temporal course of simultaneous tone-one-tone masking. *J. acoust. Soc. Am.* **81**, 1073–1077.

Moore, B.C.J. 1980a Detection cues in forward masking. In *Psychophysical, physiological and behavioural studies in hearing* (ed. G. van den Brink & F. A. Bilson). Delft University Press.

Moore, B.C.J. 1980b Mechanisms and frequency distribution of two-tone suppression in forward masking. *J. acoust. Soc. Am.* **68**, 814–824.

Moore, B.C.J. 1981 Interactions of masker bandwidth with signal duration and delay in forward masking. *J. acoust. Soc. Am.* **70**, 62–68.

Terry, M. & Moore, B.C.J. 1977 'Suppression' effects in forward masking. *J. acoust. Soc. Am.* **62**, 781–784.

B. A. WRIGHT. As Dr Moore has indicated, there is considerable evidence that the amount of forward masking can be influenced by the available listening cues. However, we list below our reasons for believing that such differences cannot easily account for our results. First, in Dr Moore's example, the masker in the reference condition was a narrowband noise, a stimulus which appears to be optimum for producing high thresholds in forward masking due to confusion about when the masker ends and the signal begins. In our experiment, the masker was a pure tone, a stimulus for which such confusion effects are considerably smaller or non-existent (Neff 1986). Second, we gated a contralateral wideband noise with the masker or suppressor to help mark the end of the masker (Moore & Glasberg 1982; Neff 1986) to reduce any small confusion effects that might have been present. Third, Dr Moore implies that thresholds are reduced in masker-plus-suppressor conditions simply because the suppressor helps to indicate the end of the masker. According to this interpretation, none of our subjects had suppression. Rather, the subjects who showed normal suppression were very good at using the timing information provided by the offset of the suppressor or the contralateral noise, and the subjects who showed negative suppression were unable to use these offset cues. Even if Dr Moore were correct, because all of our subjects had at least some signal enhancement and all showed masker enhancement, the resulting conclusion would be the same as we offered: that adaptation of psychophysical two-tone suppression cannot account for those enhancement effects.

We also think it unlikely that enhanced forward masking or masker enhancement results simply from 'distraction'. Our reasons are as follows. First, the greatest amount of masker enhancement was seen for notch widths in the enhancer of 400 Hz and greater. It is not clear how a distraction interpretation could account for a small amount of masker enhancement with a 200 Hz wide notch, and greater enhancement with wider notch widths. Second, Dr Moore suggests that the spectral change in the masker occurring 'just before' the signal distracts the subject. It would therefore seem that the amount of masker enhancement should steadily decrease as the duration of the masker was increased, and the signal onset correspondingly delayed relative to masker onset. However, in the present experiments, the amount of masker enhancement was greatest and relatively constant across the intermediate masker durations (42–102 ms). Third, confusion effects in forward masking are greatest when the masker and signal are similar in their temporal properties (Moore & Glasberg 1982; Neff 1985). One might thus predict that the greatest amount of distraction would occur when the masker and signal durations were equal. However, in the present experiments, the greatest effects were for masker durations that exceeded the 20 ms duration of the signal. Fourth, Viemeister & Bacon (1982) showed masker enhancement at signal delays greater than could be easily accounted for by confusion (Neff 1985). Fifth, we have pilot data using a signal located at the centre frequency of the notch in the enhancer (a standard masker-enhancement task) and for signals located at frequencies above and below the presumably enhanced region. According to the distraction hypothesis, masker enhancement should be obtained with all three of these signals. In our pilot experiment, thresholds were always higher in the enhanced condition than in the masker-only condition. However, these theshold differences were greater than could be accounted for by differences in the duration of the masker only for the signal centered in the notch of the enhancer. Thus, contrary to the prediction of a general distraction process, masker enhancement appears to require that the enhanced region overlap the signal in frequency.

Reference

Neff, D.L. 1985 Confusion effects with sinusoidal and narrow-band noise forward maskers. *J. acoust. Soc. Am.* **79**, 1519–1529.

[35]

Masking release for gap detection

JOSEPH W. HALL III AND JOHN H. GROSE

The Division of Otolaryngology/Head & Neck Surgery, University of North Carolina Medical School, Chapel Hill, North Carolina, U.S.A.

SUMMARY

In random noise, masking is influenced almost entirely by noise components in a narrow band around the signal frequency. However, when the noise is not random, but has a modulation pattern which is coherent across frequency, noise components relatively remote from the signal frequency can actually produce a release from masking. This masking release has been called comodulation masking release (CMR). The present research investigated whether a similar release from masking occurs in the analysis of a suprathreshold signal. Specifically, the ability to detect the presence of a temporal gap was investigated in conditions which do and do not result in CMR for detection threshold. Similar conditions were investigated for the masking level difference (a binaural masking release phenomenon). The results indicated that suprathreshold masking release for gap detection occurred for both the masking-level difference (MLD) and for CMR. However, masking release for gap detection was generally smaller than that obtained for detection threshold. The largest gap detection masking release effects obtained corresponded to relatively low levels of stimulation, where gap detection was relatively poor.

1. INTRODUCTION

The results of many auditory masking experiments using random noise maskers can be accounted for well by the critical band or auditory filter model (Fletcher 1940; Green *et al.* 1959; Margolis & Small 1975; Patterson 1976): in general, only the masking noise components within a relatively narrow frequency band around the frequency of a pure-tone signal contribute significantly to the masking of the signal. Energy relatively remote from the signal frequency has a negligible effect in comparison to noise components near the signal frequency. This relatively simple auditory filter model does not apply well, however, when the masking noise is modulated such that noise components around the signal frequency have an envelope fluctuation that is correlated with the fluctuations of noise components away from the signal frequency; such noise is referred to as comodulated. Here, the presence of comodulated distal noise components results in an improvement in detection threshold (a release from masking). This phenomenon has been termed comodulation masking release, or CMR (Buus 1985; Carlyon *et al.* 1989; Cohen & Schubert 1987; Haggard *et al.* 1990; Hall *et al.* 1984; McFadden 1986; Moore *et al.* 1990; Moore & Schooneveldt 1990; Schooneveldt & Moore 1987; Wright & McFadden 1988).

Whereas some of the conditions of the present experiment investigated CMR in a signal detection paradigm, the main focus was the extent to which CMR applies in the analysis of a suprathreshold (partially masked) signal. There are very few data currently available on this issue. What little data do exist would suggest that masking release effects may be rather small for partially masked signals. For example, whereas Grose & Hall (1992) found consistent CMR effects for the detection of speech signals in noise backgrounds, CMR effects related to the recognition of speech were small or absent. This would imply that even though speech information may be more audible in comodulated noise, the quality of that information is, in some sense, poor. Interestingly, there is some precedent for this sort of finding in another masking release paradigm, the masking-level difference, or MLD (Hirsh 1948). The MLD for detection is often defined as the difference between two thresholds. In one, the same masking noise is presented to boh ears (in phase), and the signal is also presented to both ears in phase. This is referred to as NoSo. In the other condition, the noise is again presented to the two ears in phase, but the signal is presented 180° out of phase (NoSπ). The detection threshold for NoSπ is often approximately 15 dB better than for NoSo. Even though this relatively large masking release occurs at detection threshold, Henning (1991) has found that intensity and frequency discrimination are relatively poorer for NoSπ than for NoSo stimulation, when signals are presented at low equivalent sensation level (SL) above masked threshold. Robinson & Blakeslee (1971) likewise reported that effects of binaural masking release for the ability to discriminate differences in duration were relatively small.

The present experiments were concerned with the suprathreshold task of temporal gap detection (Grose *et al.* 1989; Penner 1977; Plomp 1964; Shailer &

Phil. Trans. R. Soc. Lond. B (1992) **336**, 331–337
Printed in Great Britain

331

Moore 1983). Performance on this measure of temporal resolution was investigated for conditions giving an MLD and those giving a CMR.

2. METHODS

Three types of masking noise were used. The first type (on-signal band alone) was a 20-Hz-wide noise band centred on the signal frequency (1200 Hz). The second masker type (comodulated envelope) was composed of the 20-Hz-wide band centred on 1200 Hz, plus four additional 20-Hz-wide comodulated noise bands, centred on 400, 800, 1600 and 2000 Hz. The third masker type (random envelope) was identical to the second, except that the envelopes of the five bands were random with respect to each other. CMR for pure-tone detection threshold was measured both as the threshold difference between the case with the on-signal band alone and the case with the comodulated envelope, and as the threshold difference between the case with the random envelope and the case with the comodulated envelope. The MLD was measured only for the on-signal band alone (the threshold difference between conditions of NoSo stimulation and NoSπ stimulation). There were three parts to the experiment. The first part simply established masked pure-tone detection thresholds (0 dB sensation level), and the second and third parts investigated gap detection. Two well-practised subjects participated.

(a) Pure-tone masked detection thesholds

In the first part of the experiment, masked detection thresholds were obtained. NoSo thresholds were obtained using all three types of masking noise, to measure CMR. NoSπ thresholds were obtained only using the on-signal band alone noise. The signal was a 1200 Hz pure tone (400 ms with 50 ms cosine^2 rise fall). The stimuli were presented via Sony MDR V6 earphones. All noise stimuli were presented continuously at a pressure spectrum level of 50 dB Hz. The comodulated Gaussian noise bands were created by multiplying (Analog Devices AD534LH) a digitally synthesized complex tone composed of 5600, 5200, 4800, 4400 and 4000 Hz by a bandpass noise from 5990 Hz to 6010 Hz. Following multiplication, the upper sidebands were filtered out, leaving 20-Hz-wide bands centered on 400, 800, 1200, 1600 and 2000 Hz. These bands were recorded (using a digital audio tape recorder). The on-signal band alone was created in a similar way, but only the 4800 Hz pure tone was multiplied by the bandpass noise. The random noise was created by digitally filtering (Trinder 1983) a wideband noise source, using a sampling rate of 5.0 kHz.

(b) Gap detection thresholds for a fixed-SL pure-tone: temporal gap varied adaptively

In the second part of the experiment, the duration of a gap was varied to obtain gap detection thresholds for a signal presented at a fixed SL above masked threshold. Again, the So signal was presented in each of the three types of masking noise, and the Sπ signal was presented only in the on-signal band alone noise. In each interval of a trial, the pure tone was gated on for 1.2 s. In the target interval, the gap began 450 ms after stimulus onset and its duration was subtracted from the remaining portion of the interval to keep the total duration constant at 1.2 s. The pure-tone stimuli were gated on and off via a Wilsonics gate, set to fast (less than 10 μs) rise–fall time. To minimize spectral cues for gap detection (energy splatter), the output of the gate was digitally filtered to a 20-Hz bandwidth around the signal frequency. This procedure effectively imposes a relatively slow rise–fall time on the pure-tone signal. We measured the rise–fall time as the time taken for the amplitude of the sine wave to fall from 90% to 10% of its peak value. This time was approximately 38 ms. Gap detection thresholds were determined for masked SLs of 5, 10, 15, 20, 25 and 30 dB. It was found that for some low SL conditions, the just detectable gap was quite large (greater than 350 ms), and performance was highly variable. Data were therefore not obtained on conditions for which thresholds were not consistently smaller than 350 ms.

(c) Gap detection thresholds for a fixed-gap signal: signal SPL varied adaptively

In the third part of the experiment, the level of a pure-tone signal, having a gap of fixed duration, was varied to obtain the level at which the gap could be detected. Again, the So signal was presented in each of the three types of masking noise, and the Sπ signal was presented in the on-signal band alone. In each interval of a trial, the pure tone was gated on for 1.2 s. During the gap interval a fixed silent period was introduced. The gap began 450 ms after stimulus onset and its duration was subtracted from the remaining portion of the interval to maintain a 1.2 s total duration. The fixed gap values investigated were 25, 50 and 250 ms. Gating and digital filtering were used as above to generate gaps with minimal spectral spread.

(d) Threshold estimation procedures

All (variable-signal-level and variable-gap-duration) thresholds were determined using a three-alternative forced-choice (3AFC), three-down, one-up adaptive procedure, estimating the 79.4% detection threshold (Levitt 1971). For level variation conditions, an initial step-size of 8 dB was reduced to 4 dB after two reversals, and further reduced to 2 dB after two more reversals. A threshold run was stopped after 12 reversals, and the average of the levels at the last eight reversals was taken as the detection threshold for a run. For gap variation, the gap size was reduced or increased by a factor of 1.2 according to the three-down one-up rule. Six reversals in duration were measured and threshold was estimated as the geometric mean of the gap values at the final four reversals. The inter-stimulus interval was 300 ms. At least four threshold runs were averaged to compute the final detection threshold for a condition. Visual

Table 1. *Masked detection thresholds and the* MLDs *and* CMRs *derived from the three types of masking noise*

CMR_on-com refers to the CMR derived by subtracting the threshold in comodulated envelope noise from the threshold in the on-signal band alone; CMR_ran-com refers to the CMR derived by subtracting the threshold in comodulated envelope noise from the threshold in random envelope noise. CMRs are derived from NoSo stimuli.

| | on-signal band | | | random | comodulated | | |
	NoSo	NoSπ	MLD	NoSo	NoSo	CMR_on-com	CMR_ran-com
S1	61.5	44.3	17.2	61.7	47.7	13.8	14.0
S2	61.4	48.6	12.8	60.5	47.8	13.6	12.7
avg.	61.4	46.4	15.0	61.1	47.8	13.7	13.4

feedback was provided after each response. Stimulus presentation and response collection were controlled by an IBM AT microcomputer.

3. RESULTS AND DISCUSSION

(a) *Pure-tone masked detection thesholds*

Although the primary focus of the present experiments was on suprathreshold performance, the basic masked threshold data (table 1) will also be considered briefly. CMR is defined both in terms of the difference (in dB) between the threshold for the on-signal band alone condition and the threshold for the comodulated envelope condition, and the difference between the threshold for the random envelope condition and the threshold for the comodulated envelope condition. For NoSo, the thresholds in the on-signal band alone and random envelope noise cases were similar; thus for NoSo, the two measures of CMR were approximately the same (13–14 dB). MLD effects for the on-signal band alone are seen by comparing the NoSo thresholds to the NoSπ thresholds: the MLD was approximately 13–17 dB.

(b) *Gap detection thresholds as a function of pure-tone* SL

The gap detection thresholds as a function of the SL of the partially masked pure-tone signal are shown in figure 1. The non-masking release conditions are shown by filled symbols (circles for on-signal band, NoSo; filled triangles for random envelope, NoSo), and the masking release conditions are shown by the open symbols (open circles for the on-signal band, NoSπ; open triangles for comodulated envelope, NoSo). In the conditions where there was no masking release, gap detection improved with increasing SL, and then approached asymptote by about 15–25 dB SL. Gap detection was generally slightly better for the So on-signal band alone condition than for the So random-envelope condition. This is probably related to an effect reported by Grose & Hall (1988). In that experiment, gaps were detected in a narrow band of noise. In one condition, only the signal band was present. In the other, a second narrow band noise, which did not contain a gap, was also present. In the latter situation, the gap threshold was up to two times higher. It was hypothesized that it was the amplitude fluctuations of the noise at the distal frequency that made the gap difficult to detect at the target frequency. We have speculated (Hall & Grose 1991) that this gap detection effect is related to the modulation detection interference (MDI) phenomenon described by Yost & Sheft (1989, 1990). In the present experiment, the fluctuations of the flanking noise components in the multiband cases may have made the temporal modulation at the target frequency more difficult to detect. For this reason, of the two non-masking release

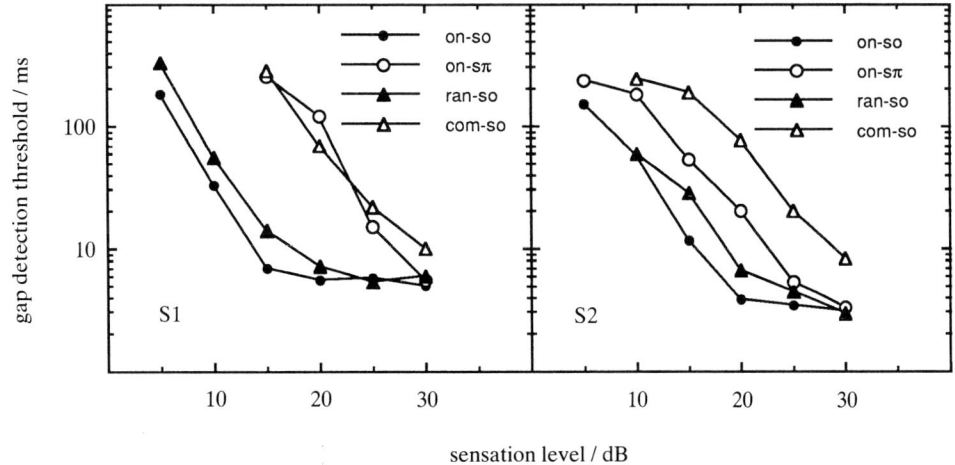

Figure 1. Gap detection thresholds (ms), plotted as a function of the SL (dB) of the partially masked pure-tone signal. On-so and on-sπ refer to NoSo and NoSπ conditions for the on-signal band alone masker; ran-so refers to the NoSo condition for the random envelope masker; com-so refers to the NoSo condition for the comodulated envelope masker.

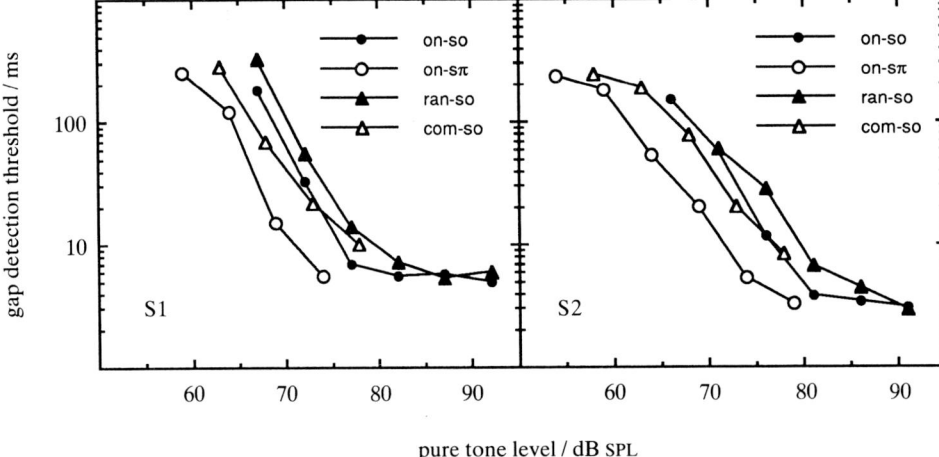

Figure 2. Gap detection thresholds (ms), plotted as a function of the SPL (dB) of the partially masked pure-tone signal. On-so and on-sπ refer to NoSo and NoSπ conditions for the on-signal band alone masker; ran-so refers to the NoSo condition for the random envelope masker; com-so refers to the NoSo condition for the comodulated envelope masker.

conditions, the So random envelope condition is probably a more appropriate control condition than the So on-signal band alone condition, when making comparisons to the multiband masking release condition.

A general result which can be seen in figure 1 is that the conditions associated with masking release (for masked signal detection) resulted in relatively poor gap detection thresholds at low SLs. For example, S1 did not demonstrate a reliable gap detection threshold for SLs less than 15 dB for the So signal in comodulated noise, and Sπ signals in the on-signal band alone. S2 also showed inferior performance for the masking release conditions, although this subject was able to achieve stable performance at low SLs in all four conditions. The results of the two subjects, considered in terms of SL above masked detection threshold, indicate that gap detection is relatively poor in conditions associated with MLDs or CMRs.

The data in figure 1 may be somewhat misleading in terms of the relative advantage or disadvantage for gap detection in the masking release conditions. Although the data in figure 1 show clearly that gap detection is better in the non-masking release conditions than in the masking release conditions at low SL, it should be remembered that the signal is at a higher signal-to-noise ratio in the non-masking release conditions than in the masking release conditions, given a fixed SL. It is also possible to consider the result in terms of the SPL of the partially masked pure-tone signal. Figure 2 plots the gap detection results as a function of the SPL of the pure-tone signal (the data are the same as in figure 1, but plotted in terms of pure-tone signal SPL rather than SL above masked threshold). The data plotted in this way show that, in general, stable gap detection performance was obtained at lower SPLs for the masking release conditions than for the non-masking release conditions. Comparisons between the filled and open circles provide an indication of the MLD effect, and comparisons between the filled and open triangles provide an

indication of the CMR effect. These comparisons indicate that masking release (MLD and CMR) did occur in the suprathreshold gap detection task. However, the masking release is smaller than that associated with masked detection threshold (table 1). Some of the data in figure 2 suggest that masking release the gap detection may be generally greater at low SPLs, where gap detection in general is relatively poor, than at higher SPLs. The question of the magnitude of the masking release for suprathreshold gap detection is addressed more precisely by considering the data from conditions where the level of the signal was varied to obtain threshold for the detection of a fixed-duration gap.

(c) Gap detection thresholds as a function of fixed-gap duration

Figure 3 shows signal levels at which subjects could detect the signal containing a gap, as a function of the fixed gap duration. The figures indicate that the conditions resulting in masking release at detection threshold generally resulted in some degree of masking release for the detection of a gap in a pure-tone signal. Table 2 shows the signal level at which gap threshold was obtained for NoSo and NoSπ conditions, as well as derived MLDs and CMRs. Even though CMRs are shown using both the on-signal band alone baseline and the random envelope baseline, the random envelope baseline is probably the more appropriate (see discussion above), and only these CMRs will be addressed. NoSo CMRs for gap detection (Table 2) were typically rather small, except for the 250 ms gap, where the CMR was from 7–9 dB. Even this CMR was smaller than that obtained at detection threshold (table 1). S1's MLDs for the detection of temporal gaps were considerably smaller than those obtained for masked detection threshold. S2's gap detection MLD was small for the shortest gap, but her MLDs were sizeable for the 50 and 250 ms gaps (10.4 and 11.4 dB, respectively). Again, there was a trend for the gap

Table 2. *Thresholds at which gaps could be detected and gap detection MLDs and CMRs derived from the three types of masking noise*

CMR$_{on-com}$ refers to the CMR derived by subtracting the threshold in comodulated envelope noise from the threshold in the on-signal band alone; CMR$_{ran-com}$ refers to the CMR derived by subtracting the threshold in comodulated envelope noise from the threshold in random envelope noise. CMRs are derived from NoSo stimuli.

	on-signal band			random	comodulated		
	NoSo	NoSπ	MLD	NoSo	NoSo	CMR$_{on-com}$	CMR$_{ran-com}$
25 ms gap							
S1	72.1	67.8	4.3	73.2	70.2	1.9	3.0
S2	70.0	62.6	7.4	70.3	65.9	4.1	4.4
avg.	71.0	55.2	5.8	71.8	68.0	3.0	3.8
50 ms gap							
S1	69.6	65.9	3.7	71.9	67.3	2.3	4.6
S2	68.9	58.5	10.4	69.4	65.8	3.1	3.6
avg.	69.2	62.2	7.0	70.6	66.6	2.7	4.1
250 ms gap							
S1	66.1	60.2	5.9	69.6	60.3	5.8	9.3
S2	66.2	54.8	11.4	66.9	59.6	6.6	7.3
avg.	66.2	57.5	8.7	68.2	60.0	6.2	8.3

detection MLD to increase slightly with increasing value of the gap.

4. GENERAL DISCUSSION

The present study showed two main effects: (i) at equal, low SL, gap detection performance was better under non-masking release conditions than masking release conditions; (ii) at equal signal SPLs, both MLD and CMR masking release occurred, but the magnitude of the effects was generally less than obtained for detection threshold. As regards the MLD, these results are generally consistent with previous investigations of suprathreshold signal analysis (Gebhardt *et al.* 1971; Henning 1991; Robinson & Blakeslee 1971; Townsend & Goldstein 1971).

As noted above, CMRs (and, perhaps, MLDs) for gap detection were often greater for the 250 ms gap than for the shorter gap durations. There are at least two factors that may account for this result as regards CMR. One is related to the fact that previous research has shown that the information contributing to CMR occurs in the dip regions of the noise (Grose & Hall 1989). In gap conditions where the gap is short, the probability that part of the gap will occur in a dip region of the masker will be relatively low. However, when the gap is long, the probability will be greater that part of the gap will occur in a dip region. An across-frequency analysis in such a dip region would indicate a good correlation of envelope at the signal frequency with envelopes at flanking frequencies. This could cue the presence of a gap in the pure-tone signal (when the gap is not present, the signal will cause the envelope at the signal frequency to differ from that at the flanking

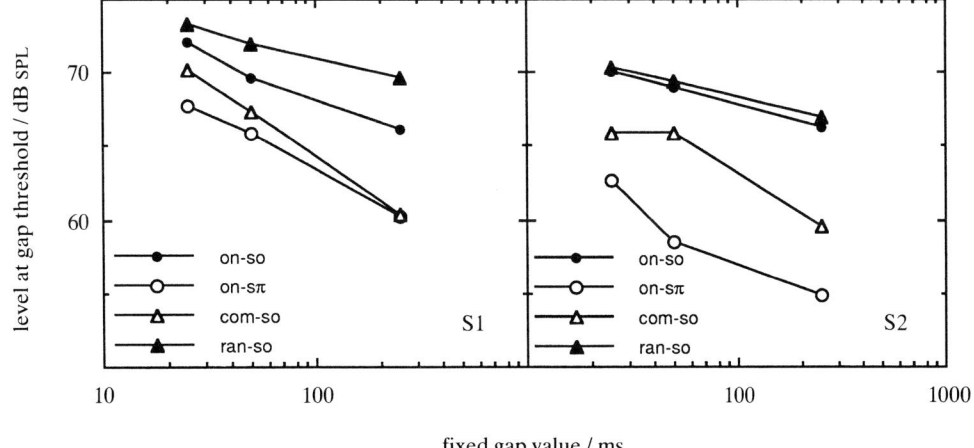

Figure 3. SPL (dB) at gap threshold, plotted as a function of the fixed gap value (ms) of the pure-tone signal. On-so and on-sπ refer to NoSo and NoSπ conditions for the on-signal band alone masker; ran-so refers to the NoSo condition for the random envelope masker; com-so refers to the NoSo condition for the comodulated envelope masker.

frequencies, and the envelope correlation will be relatively poor). The second reason that CMR might be greater at large gap values applies also to the MLD. This is simply that when the level of the signal is sufficiently high, performance will be near optimal both for masking release and non-masking release conditions; that is, at a high enough signal-to-noise ratio, the masking noise becomes essentially inconsequential. In this respect, masking release effects must be confined to relatively low signal-to-noise ratios. One way of interpreting the data of the present experiment is that masking release for gap detection is relatively large at signal-to-noise ratios where gap detection is poor (e.g. the 250 ms gap). That is, masking release occurs for suprathreshold signals, but the masking release is large only when the precision of temporal resolution called for is relatively coarse. When the temporal resolution must be relatively precise (e.g. 25 ms or better), masking release is likely to be small.

There are two caveats that should be mentioned regarding the present study. One is that the conclusions reached here are based upon the results of only two subjects and a limited number of test conditions. In this sense, the present results are preliminary. The second caveat concerns the MLD. The gap detection MLD effects for suprathreshold signals obtained here are comparable to the trends that have been reported for other measures of suprathreshold binaural masking release: that is, suprathreshold masking release effects are generally smaller than found for detection threshold. However, previous work has used low-frequency signals where the MLD is relatively large, and presumably based primarily upon cues of interaural time difference. The present study used a 1200 Hz pure tone signal. Even though the MLD is relatively large for this frequency when the noise bandwidth is narrow, the stimulus envelope cues are probably more important at the 1200 Hz frequency than at the lower frequencies that have been used in previous studies. It is possible that there are important differences between suprathreshold MLDs that are based upon fine structure time cues and suprathreshold MLDs based upon envelope cues. Comparisons between the present MLD results and the MLD results of studies that have used low-frequency signals should therefore be made with caution.

This research was supported by a grant from the U.S. Air Force Office of Scientific Research.

REFERENCES

Buus, A. 1985 Release from masking caused by envelope fluctuations. *J. acoust. Soc. Am.* **78**, 1958–1965.

Carlyon, R.P., Buus, S. & Florentine, M. 1989 Comodulation masking release for three types of modulator as a function of modulation rate. *Hear. Res.* **42**, 37–46.

Cohen, M.F. & Schubert, E.D. 1987 Influence of place synchrony on detection of a sinusoid. *J. acoust. Soc. Am.* **81**, 452–458.

Fletcher, H. 1940 Auditory patterns. *Rev. Mod. Phys.* **12**, 47–65.

Gebhardt, C.J., Goldstein, D.P. & Robertson, R.M. 1971 Frequency discrimination and the MLD. *J. acoust. Soc. Am.* **51**, 1228–1232.

Green, D.M., McKey, M.J. & Licklider, J.C.R. 1959 Detection of a pulsed sinusoid in noise as a function of frequency. *J. acoust. Soc. Am.* **31**, 1446–1452.

Grose, J.H., Eddins, D. & Hall, J.W. 1989 Gap detection as a function of stimulus bandwidth with fixed high-frequency cutoff in normal-hearing and hearing-impaired listeners. *J. acoust. Soc. Am.* **86**, 1747–1755.

Grose, J.H. & Hall, J.W. 1988 Gap detection in a narrowband noise with either a comodulated or noncomodulated flanking band. *J. acoust. Soc. Am.* **83**, S34.

Grose, J.H. & Hall, J.W. 1989 Comodulation masking release using SAM tonal complex maskers: Effects of modulation depth and signal position. *J. acoust. Soc. Am.* **85**, 1276–1284.

Grose, J.H. & Hall, J.W. 1992 Comodulation masking release for speech stimuli. *J. acoust. Soc. Am.* (Submitted.)

Haggard, M.P., Hall, J.W. & Grose, J.H. 1990 Comodulation masking release as a function of bandwidth and test frequency. *J. acoust. Soc. Am.* **87**, 269–283.

Hall, J.W. & Grose, J.H. 1991 Some effects of auditory grouping factors on modulation detection interference (MDI). *J. acoust. Soc. Am.* 3028–3036.

Hall, J.W., Haggard, M.P. & Fernandes, M.A. 1984 Detection in noise by spectrotemporal pattern analysis. *J. acoust. Soc. Am.* **76**, 50–56.

Henning, G.B. 1991 Frequency discrimination, amplitude discrimination, and the binaural masking-level difference: Some anomalous results. In *Auditory physiology and perception* (ed. Y. Cazals, L. Demany & K. Horner), pp. 507–512. Oxford: Pergamon Press.

Hirsh, I.J. 1948 Binaural summation and interaural inhibition as a function of the level of the masking noise. *J. acoust. Soc. Am.* **20**, 205–213.

Levitt, H. 1971 Transformed up-down methods in psychoacoustics. *J. acoust. Soc. Am.* **49**, 467–477.

Margolis, R.H. & Small, A.M. 1975 The measurement of critical masking bands. *J. Speech Hear. Res.* **40**, 1414–1419.

McFadden, D. 1986 Comodulation masking release: Effects of varying the level, duration, and time delay of the cue band. *J. acoust. Soc. Am.* **80**, 1658–1667.

Moore, B.C.J., Hall, J.W., Grose, J.H. & Schoonevelt, G.P. 1990 Some factors affecting the magnitude of comodulation masking release. *J. acoust. Soc. Am.* **88**, 1694–1702.

Moore, B.C.J. & Schooneveldt, G.P. 1990 Comodulation masking release as a function of bandwidth and time delay between on-frequency and flanking-band maskers. *J. acoust. Soc. Am.* **88**, 725–731.

Patterson, R.D. 1976 Auditory filter shapes derived with noise stimuli. *J. acoust. Soc. Am.* **59**, 640–654.

Penner, M.J. 1977 Detection of temporal gaps in noise as a measure of the decay of auditory sensation. *J. acoust. Soc. Am.* **61**, 552–557.

Plomp, R. 1964 Rate of decay of auditory sensation. *J. acoust. Soc. Am.* **36**, 277–282.

Robinson, D.E. & Blakeslee, E.A. 1971 Comparison of detection and discrimination of tones of different durations under homophasic and antiphasic conditions. *J. acoust. Soc. Am.* **49**, 102.

Schooneveldt, G.P. & Moore, B.C.J. 1987 Comodulation masking release (CMR): Effects of signal frequency, flanking band frequency, masker bandwidth, flanking band level, and monotic versus dichotic presentation of the flanking band. *J. acoust. Soc. Am.* **82**, 1944–1956.

Shailer, M.J. & Moore, B.C.J. 1983 Gap detection as a function of frequency, bandwidth and level. *J. acoust. Soc. Am.* **74**, 467–473.

Townsend, T.H. & Goldstein, D.P. 1971 Suprathreshold binaural unmasking. *J. acoust. Soc. Am.* **51**, 621–624.

Trinder, J.R. 1983 Hardware-software configuration for high performance digital filtering in real time. *Proc. IEEE Int. Conf. Acoust. Speech Signal Process.* **2**, 687–690.

Wright, B.A. & McFadden, D. 1988 Comodulation masking release with delayed signal onsets. *J. acoust. Soc. Am.* **83**, S34.

Yost, W.A. & Sheft, S. 1989 Across-critical-band processing of amplitude-modulated tones. *J. acoust. Soc. Am.* **85**, 848–857.

Yost, W.A. & Sheft, S. 1990 A Comparison among three measures of cross-spectral processing of amplitude modulation with tonal signals. *J. acoust. Soc. Am.* **87**, 897–900.

Modulation discrimination interference and auditory grouping

BRIAN C. J. MOORE AND MICHAEL J. SHAILER

Department of Experimental Psychology, University of Cambridge, Downing Street, Cambridge CB2 3EB, U.K.

SUMMARY

The detection of a change in the modulation pattern of a (target) carrier frequency, f_c (for example a change in the depth of amplitude or frequency modulation, AM or FM) can be adversely affected by the presence of other modulated sounds (maskers) at frequencies remote from f_c, an effect called modulation discrimination interference (MDI). MDI cannot be explained in terms of interaction of the sounds in the peripheral auditory system. It may result partly from a tendency for sounds which are modulated in a similar way to be perceptually 'grouped', i.e. heard as a single sound. To test this idea, MDI for the detection of a change in AM depth was measured as a function of stimulus variables known to affect perceptual grouping, namely overall duration and onset and offset asynchrony between the masking and target sounds. In parallel experiments, subjects were presented with a series of pairs of sounds, the target alone and the target with maskers, and were asked to rate how clearly the modulation of the target could be heard in the complex mixture. The results suggest that two factors contribute to MDI. One factor is difficulty in hearing a pitch corresponding to the target frequency. This factor appears to be strongly affected by perceptual grouping. Its effects can be reduced or abolished by asynchronous gating of the target and masker. The second factor is a specific difficulty in hearing the modulation of the target, or in distinguishing that modulation from the modulation of other sounds that are present. This factor has effects even under conditions promoting perceptual segregation of the target and masker.

1. INTRODUCTION

It is widely assumed that the peripheral auditory system contains a bank overlapping bandpass filters (the auditory filters) (Fletcher 1940). When an observer is trying to detect a narrowband signal in the presence of a masking sound, it is usually assumed that performance is based on the output of the single auditory filter which gives the highest signal-to-masker ratio (Fletcher 1940). Although this model works well in many situations (Patterson & Moore 1986) it clearly fails in others. In the situations considered in this paper, the outputs of auditory filters tuned away from the signal frequency degrade signal detection. It appears that sometimes subjects cannot attend to a single auditory filter even though it would pay them to do so. This degradation seems to happen mainly when the masker and signal are modulated in some way and the task of the subject is to detect a change in the modulation of the signal (Yost & Sheft 1989; Yost *et al.* 1989; Moore *et al.* 1990, 1991; Wilson *et al.* 1990). Hence, this phenomenon has been called modulation detection interference (Yost & Sheft 1989) or modulation discrimination interference (Moore & Jorasz 1992) (MDI).

As an example of MDI, we will briefly describe some experiments reported by Moore *et al.* (1991), as the experiments described in this paper use very similar stimuli. Moore *et al.* determined how thresholds for

detecting an increase in modulation depth (sinusoidal amplitude modulation, AM, or frequency modulation, FM) of a 1000 Hz carrier frequency (the target) were affected by the presence of modulated carriers (maskers) with frequencies of 230 Hz and 3300 Hz. The carrier frequencies of the maskers were sufficiently far from the target frequency that they would have produced a negligible output from the auditory filter centred at the target frequency. The target was presented twice for a duration of 1000 ms, with 300 ms inter-stimulus interval, and the subject was required to indicate the interval in which the sound was more modulated. The maskers were gated with the target. They found that modulation increment thresholds were increased (worsened) in the presence of the maskers. This MDI effect was greatest when the target and maskers were modulated at similar rates, but the effect was broadly tuned for modulation rate.

2. THE POSSIBLE ROLE OF PERCEPTUAL GROUPING IN MDI

In everyday life we often listen to several sound sources simultaneously. The auditory system has to decide which 'elements' of the complex mixture arise from one source, and which from another. The process of doing this is described as 'perceptual grouping' or 'stream formation'; the elements of the sound are grouped across-frequency and across time to form

Phil. Trans. R. Soc. Lond. B (1992) **336**, 339–346
Printed in Great Britain

339

© 1992 The Royal Society and the authors

[45]

percepts of coherent streams each with its own loudness, pitch, timbre and location (Moore 1989; Bregman 1990). One principle that operates in perceptual grouping is 'common fate'; elements of a sound that change in the same way tend to be grouped and heard as a single stream. Two applications of this principle are relevant here: elements that start and stop together tend to be perceived as a single sound; and elements that are modulated in a similar way tend to be perceived as a single sound.

Yost & Sheft (1989) suggested that MDI might be a consequence of perceptual grouping. In the case of AM sounds, the common AM of the target and maskers might make them fuse perceptually (McAdams 1984; Bregman *et al.* 1990), making it difficult to 'hear out' the modulation of the target sound. This explanation is consistent with the subjective impression of subjects in these experiments. However, certain aspects of the results on MDI are difficult to reconcile with an explanation in terms of perceptual grouping. One would expect that widely spaced frequency components would only be grouped perceptually if their modulation pattern was very similar. Grouping would not be expected, for example, if the components were modulated at different rates. Yet, it is possible to obtain large amounts of MDI under conditions where the target and the maskers are modulated at different rates (Yost *et al.* 1989; Wilson *et al.* 1990; Moore *et al.* 1991).

The present experiments were intended to clarify the role of perceptual grouping in MDI. The first experiment was similar to that of Moore *et al.* (1991). Thresholds for detecting a change in AM depth of a target sound were measured with and without maskers whose carrier frequencies were remote from the target frequency. However, in this experiment we manipulated a second factor that is known to affect perceptual grouping, namely the overall duration of the stimuli. Sounds that are gated on and off synchronously for a brief duration tend to fuse perceptually and be heard as a single sound, whereas at longer durations more than one sound may be perceived (Moore *et al.* 1986). If perceptual grouping is responsible for MDI, then at short durations the MDI produced by the synchronous gating of the target and maskers would be expected to combine with the MDI produced by modulation of the maskers to produce a greater overall effect. The second experiment used the same subjects and similar stimuli, but involved a rating task intended to provide a direct indication of the extent to which grouping of the target with the masker made it difficult to hear the modulation of the target sound. Because we wish to compare the results of the two experiments, the method for each will be described first.

3. METHOD

(a) *Experiment 1*

Thresholds were measured for detecting an increase in modulation depth of an AM target. The target was presented either alone, or with various additional sounds.

(i) *Stimuli*

The target was a 1000 Hz carrier amplitude modulated at a 10 Hz rate. The phase of the modulation was random relative to the onset of the stimulus. The masker, when present, consisted of two carriers, chosen to be non-harmonically related to the target. The two carriers were always modulated at the same depth as each other, and at the same rate. Their modulation index, m, was either 0 (no modulation) or 0.5. They were centred at 230 Hz and at 3300 Hz, the same as used by Moore *et al.* (1991). When m was 0.5, the modulation rate of the masker was either 4, 10 or 25 Hz. When the rate was 10 Hz, the masker was modulated either in phase with the target or in antiphase. Two conditions were also included where the masker consisted of a single carrier frequency, either 230 Hz or 3300 Hz, modulated at 10 Hz in phase with the target.

All carriers were presented at a level of 60 dB SPL. The root-mean-square pressure was held constant regardless of modulation depth. Thus, intensity changes in the target could not be used as a cue for detecting changes in modulation depth.

On each trial, two stimuli were presented, separated by a silent interval of 300 ms. Each stimulus had 50 ms raised-cosine rise and fall ramps. The steady state duration was either 100, 300 or 1500 ms, giving overall durations of 200, 400 or 1600 ms. In one stimulus the 1000 Hz carrier was modulated with index $m = 0.25$. In the other, m was greater. The order of the two stimuli was random.

All stimuli were generated using a Masscomp 5400 computer equipped with 16-bit digital-to-analog converters. The sampling rate was 10 kHz, and stimuli were lowpass filtered at 4 kHz (-3 dB) using a Fern Electronics EF16 filter with an attenuation rate of 100 dB per octave. Subjects were tested individually in a double-walled sound attenuating chamber. Stimuli were delivered via a manual attenuator (Hatfield 2125) to one earpiece of a Sennheiser HD414 headset.

(ii) *Procedure*

An adaptive 2AFC procedure was used to estimate the 79.4% correct point on the psychometric function. A run always started with a large change in m in the signal interval. After three successive correct responses the change in m was reduced whereas after each incorrect response it was increased. The value of m was not allowed to be greater than 1.0. Initially, the step size for the change in m was 5 dB in units of $20\log(m)$. After four reversals, the step size was decreased to 2 dB and eight further reversals were obtained. The mean value of $20\log(m)$ at the last eight reversals was used to estimate the change in m (Δm) corresponding to threshold. At least four estimates were obtained for each condition, and the threshold was calculated as the geometric mean of the four. When the standard deviation of the log values exceeded 0.2, at least one further estimate was obtained and all estimates were averaged. The standard deviation of the log values was typically between 0.05 and 0.15. The standard error of the log values was never greater than 0.1 and was typically about

0.05. Correct-answer feedback was given after each trial by means of lights on the response box.

During initial training, performance with the target sound alone stabilized quite rapidly, but thresholds continued to decrease for some time in the conditions with maskers; all subjects found these conditions very difficult at first, and they sometimes had difficulty scoring above chance. All subjects were given at least 15 h practice before collection of the data reported here. However, even after this time there was some evidence of systematic improvements in certain conditions, especially those involving two modulated maskers. Subjects were given several additional hours of practice in these conditions, until their performance appeared to be stable.

(iii) Subjects

Four subjects with normal hearing at all audiometric frequencies were used. One was author M.S. The others were paid for their services.

(b) Experiment 2

Subjects were presented with the target (a sinusoid amplitude modulated at a 10 Hz rate with $m = 0.25$, and with a carrier frequency close to 1000 Hz) followed by the target together with some maskers. They were asked to listen to the modulation of the target alone, and to rate (on a scale from 1 to 7) how clearly they could hear the modulation of the target in the complex mixture. To prevent subjects remembering the pitch of the target across trials, its carrier frequency was randomly varied ($\pm 10\%$, rectangular distribution) from trial to trial. The maskers were the same as used in experiment 1. The first and second sounds always had the same overall duration of either 200, 400 or 1600 ms, and the interval between the two stimuli in a trial was 300 ms. To control for possible range effects, all stimuli (types of masker and durations) were presented in quasi-random order within a single large block of trials. The only constraint was that each stimulus should be presented five times during the first half of a block and five times during the second half. Each block took about 15–20 min. Three blocks were run for each subject. The first two ratings of each stimulus in a block tended to be rather variable, and also to be inconsistent with ratings obtained later on in that block, so they were discarded; data presented are based on 24 ratings per stimulus.

4. RESULTS

Some individual differences were apparent in the results, but the overall pattern was similar across subjects. For simplicity, we present only the mean results. The upper panels of figure 1 show the results for experiment 1, while the lower panels show the corresponding results for experiment 2. Open hexagons show results without any masker. Error bars indicate ± one standard error of the mean across subjects; they thus give an idea of the degree of individual variability. The likely error of the data

points is somewhat less than indicated by the error bars. The rating data are plotted with values decreasing up the ordinate to make it easier to see similarities and differences between the results of the two experiments; one would expect large amounts of MDI to be associated with low clarity ratings.

Consider first the results shown in the left-hand panels. MDI is usually described as a difficulty in hearing modulation at one frequency when modulation is also present at another frequency. Consistent with this, most previous researchers have not found interference effects for unmodulated maskers at relatively long durations (Yost & Sheft 1989; Moore *et al.* 1991). However, our results show some interference for the unmodulated maskers. The effect varied somewhat with duration; at 1600 ms, thresholds increased (relative to those with no masker) by a factor of 1.8, whereas at 200 ms they increased by a factor of 2.3. This is broadly consistent with an explanation for the interference effect in terms of perceptual grouping. These findings may indicate a need to broaden the definition of MDI to include interference in modulation discrimination caused by unmodulated maskers.

Modulated maskers produced more MDI. At the longest duration, maskers modulated in phase with the target produced more MDI than maskers modulated in antiphase. However, at the two shorter durations the phase effect was very small. Using similar stimuli, Moore *et al.* (1991) found no significant effect of modulator phase for a duration of 1000 ms, although one of their subjects did show some evidence of a phase effect.

As expected, clarity ratings (bottom left panel) were highest when no maskers were present. The ratings decreased when maskers were added. The effect of duration was small. Ratings were not generally lower when the maskers were modulated than when they were unmodulated. Thus, there is a discrepancy between the results of the two experiments; modulated maskers produced the greatest MDI, but they did not have the greatest effect on subjective clarity of the modulation.

The middle panels illustrate the effect of the modulation rate of the maskers, which always had carriers of 230 and 3300 Hz. Data for the 10 Hz rate are averaged over the in phase and antiphase conditions. Moore *et al.* (1990) found that MDI measured using similar stimuli was tuned for modulation rate, peaking at 10 Hz, but the tuning was very broad. The data in figure 1 do not show any clear tuning; MDI did not vary greatly with modulation rate of the maskers. This may reflect the fact that the range of modulation rates (4–25 Hz) was smaller than used by Moore *et al.* (1991) (2–50 Hz). Clarity ratings were somewhat lower when the maskers were modulated at 4 Hz or 25 Hz than when they were modulated at 10 Hz. Thus, once again, there is a discrepancy between the results of the two experiments.

The right-hand panels show the effect of the number of masker carriers. There is a clear trend for two carriers to produce more MDI than one. In this case, there is a reasonable correspondence between the

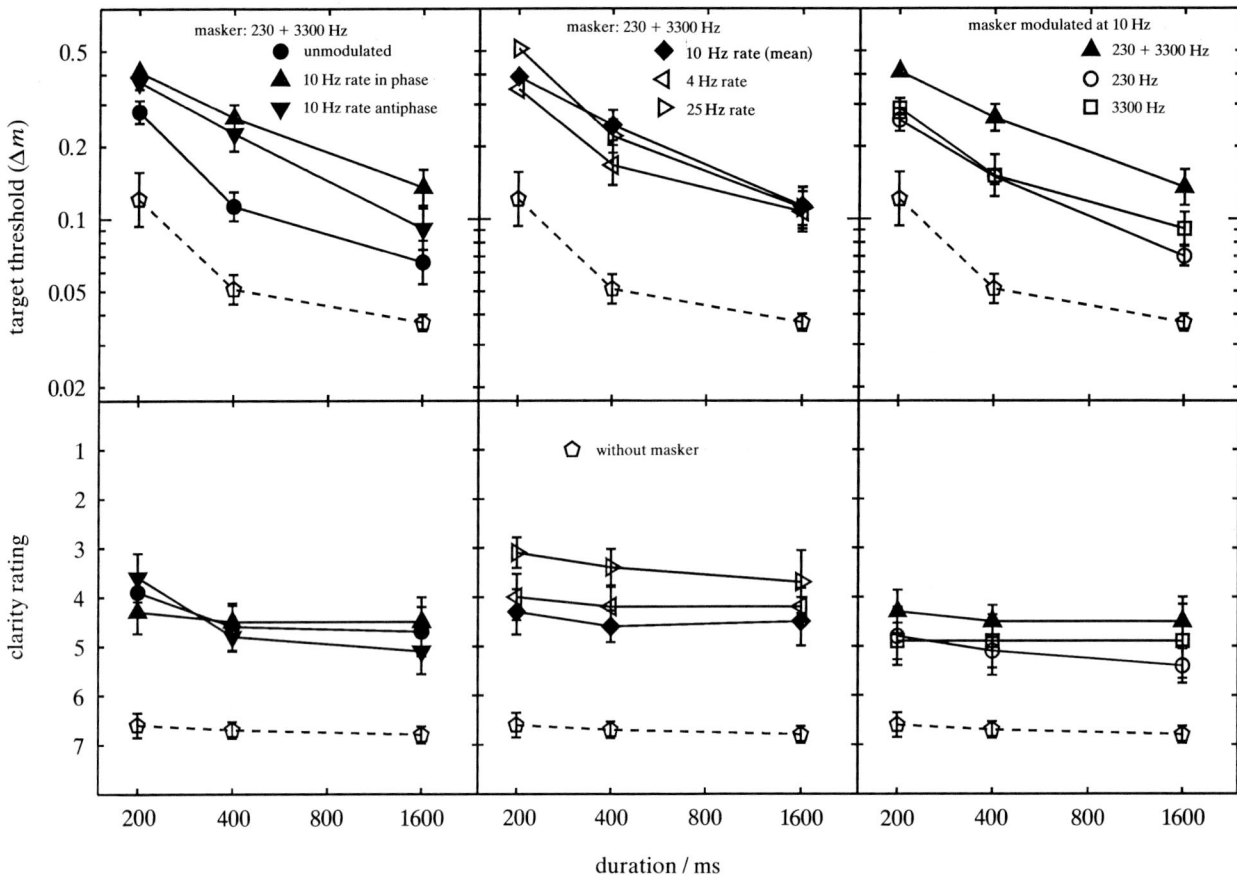

Figure 1. Results of experiment 1 (upper panels) and experiment 2 (lower panels). Results are means across four subjects. Error bars indicate ± one standard error of these means. Error bars are not shown where they would be smaller than the symbol used. The masker type is indicated in the figure.

clarity ratings and the thresholds. Clarity ratings were lower for two interfering carriers than for one.

In all conditions of experiment 1, thresholds declined with increasing duration. However, the question of whether MDI declines with increasing duration is somewhat unclear; it depends on how 'amount' of MDI is defined. If the thresholds are expressed relative to the thresholds obtained without any maskers, there is no clear evidence of a change in MDI with duration. If perceptual grouping were entirely responsible for the MDI produced by modulated maskers, then at short durations the MDI produced by the synchronous gating of the target and maskers would have been expected to combine with the MDI produced by modulation of the maskers to produce a greater overall effect. This did not happen.

In summary, the results of experiment 1 suggest that some factor other than perceptual grouping plays a role in MDI. This conclusion is supported by the finding that maskers giving the largest amount of MDI did not always produce the lowest clarity ratings.

5. EXPERIMENTS 3 AND 4

In experiments 3 and 4 we investigated the effect of another factor known to influence perceptual grouping, namely temporal asynchrony between the target sound and the masker. Asynchrony has been shown to have a powerful tendency to produce perceptual

segregation (Rasch 1978; Darwin 1981; Darwin & Sutherland 1984; Roberts & Moore 1991). Thus, we expected that the asynchrony might counteract the effects of perceptual grouping produced by the common modulation of the target and maskers, reducing MDI. Experiment 3 was similar to experiment 1; thresholds were measured for detecting a change in modulation depth of the target. Experiment 4 was a rating experiment, like experiment 2, but using stimuli similar to those used in experiment 3.

(a) Conditions

The target always had an overall duration of 400 ms. In one set of conditions, the masker was gated on before the target, with an onset asynchrony of 0, 25, 50, 100 or 200 ms. The masker and target ended simultaneously. In a second set of conditions, the target and masker started simultaneously, but the masker continued after the target for 0, 25, 50, 100 or 200 ms. The masker was either absent, unmodulated, or modulated at a 10 Hz rate either in phase with the target or in antiphase. The masker consisted of two carriers, one at 230 Hz and the other at 3300 Hz. Other aspects of the stimuli were the same as for experiments 1 and 2.

(b) Procedure and subjects

The procedure was essentially the same as for

experiments 1 and 2. In experiment 4, two separate blocks of trials were conducted, one using conditions where the masker was gated on before the target, and the other using conditions where the masker was gated off after the target. Data presented are based on a minimum of 20 ratings per stimulus per subject. Three of the subjects from experiments 1 and 2 were used, including author M.S.

6. RESULTS

The results were broadly similar across subjects, so only mean results will be presented. The upper panels of figure 2 show data for experiment 3, and the lower panels show data for experiment 4. Both when the masker was gated on before the target (top left panel) and when it was gated off after the end of the target (top right panel), MDI decreased with increasing asynchrony, but the effect was somewhat greater for onset than for offset asynchronies. For the unmodulated masker, the amount of MDI was almost absent at the largest asynchrony used. This is consistent with the idea that the MDI produced by the unmodulated masker in experiment 1 occurred because the synchronous gating of the target and masker caused them to be grouped perceptually. The asynchrony in experiment 3 apparently was sufficient to create perceptual

segregation of the target and masker, thus eliminating the MDI.

The amount of MDI also declined with increasing asynchrony for the modulated maskers. However, even at the longest asynchrony, MDI still occurred. The results of Hall & Grose (1991) and of Moore & Jorasz (1992) are consistent with this. Moore & Jorasz found that an onset asynchrony of 500 ms between the masker and target was not sufficient to abolish MDI when the masking sounds were modulated at the same rate as the target. Previous work examining the effects of asynchrony on perceptual grouping (Rasch 1978; Darwin 1981; Darwin & Sutherland 1984; Bregman 1990; Roberts & Moore 1991) indicates that an asynchrony of 200 ms is usually sufficient to produce perceptual segregation. Thus, our finding that MDI persisted at an asynchrony of 200 ms suggests that factors other than auditory grouping play a role in MDI.

The results of experiment 4 (bottom panels) show a very different pattern. Surprisingly, there was no clear change in clarity ratings with either onset or offset asynchrony. If anything, there was a slight trend for clarity ratings to decrease with increasing onset asynchrony. There is another discrepancy between the threshold data and the rating data: the unmodulated masker produced less MDI but gave lower clarity ratings than the modulated maskers.

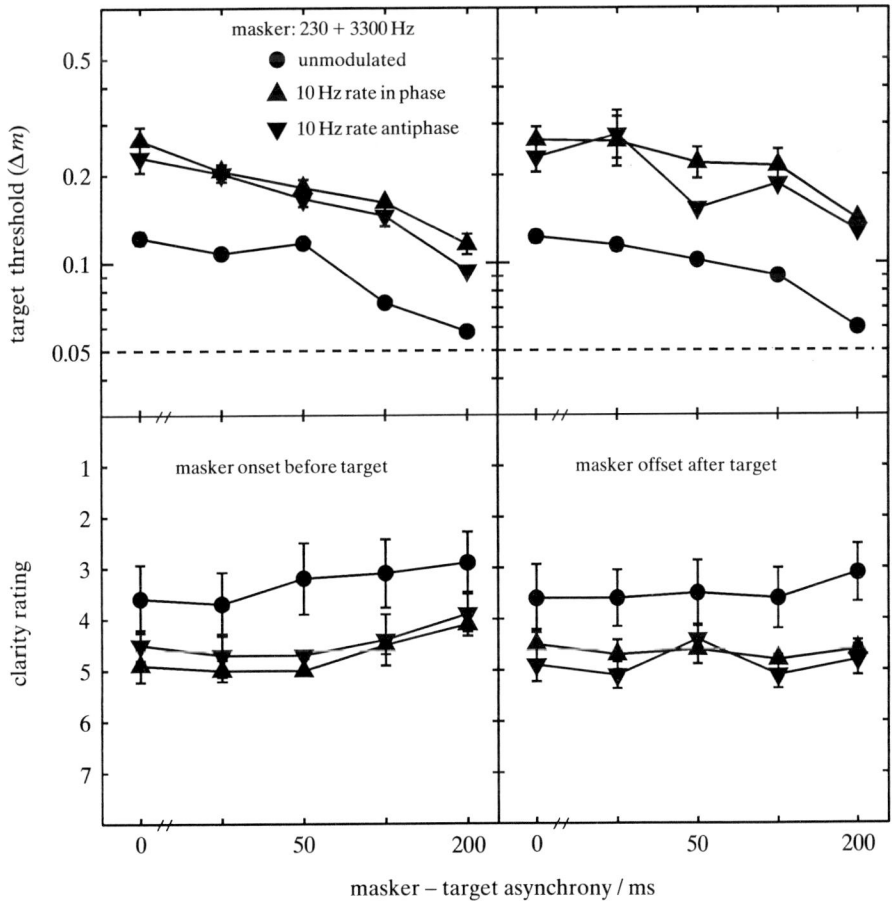

Figure 2. Results of experiment 3 (upper panels) and experiment 4 (lower panels). The dashed line indicates the threshold obtained without any masker. Otherwise as figure 1.

7. GENERAL DISCUSSION

Some aspects of the results are consistent with the idea that perceptual grouping plays a role in MDI. In experiment 1, the synchronous gating of the target and masker was sufficient to produce some MDI even when the masker was unmodulated. Experiment 3 showed that this MDI was largely abolished by gating the masker on 200 ms before the target or off 200 ms after the target, conditions designed to produce perceptual segregation of the masker and target.

On the other hand, several aspects of the results suggest that perceptual grouping is not the only factor involved. Firstly, experiment 1 showed that the MDI produced by modulated maskers was very broadly tuned for modulation rate. As pointed out by Moore *et al.* (1991), one would not expect perceptual grouping of components whose modulation frequencies were very different. Secondly, there were several discrepancies between the results of the masking experiments and of the rating experiments. The conditions giving the largest MDI did not always give the lowest clarity ratings. Finally, the MDI produced by modulated maskers was not abolished by an onset or offset asynchrony of 200 ms, even though that asynchrony was probably large enough to produce clear perceptual separation of the target and masker.

In interpreting the role of perceptual grouping in MDI, a distinction should be made between the ability to hear a pitch corresponding to the target frequency, and the ability to hear the modulation of the target frequency. It seems clear that the pitch of a given component can be harder to hear when that component is grouped perceptually with other sounds (Bregman & Pinker 1978; Moore *et al.* 1986; Bregman 1990; Roberts & Bregman 1991). Thus MDI might be a consequence of the failure to hear out the target frequency. If the target frequency could not be heard, then it would be difficult to judge the modulation of the target separately from the modulation of the other sounds present.

When we started the rating experiments, we had the above reasoning in mind. Hence, initially we asked the subjects to rate how clearly they could hear the target frequency in the complex mixture of sounds. To our surprise, the ratings were uniformly very high (all sixes and sevens), regardless of the masker type. It appeared that subjects had no difficulty in hearing out the target frequency. Hence, the instructions were modified; the rating data presented were obtained by asking subjects to rate how clearly they heard the modulation of the target in the complex mixture.

Given that the target frequency could be clearly heard in the rating experiment, even when the target and masker were gated synchronously, the results of experiment 4 become less surprising. The onset and offset asynchrony presumably had little effect on the clarity ratings because the target frequency was already clearly audible. This, however, creates a further puzzle; why did MDI change with asynchrony when the clarity ratings did not?

A possible explanation is that the stimulus sequence used in the rating experiments – target tone followed by target plus maskers – served to focus attention on the frequency region of the target, making it easier to hear out the target frequency. To check on this, experiment 2 was partially re-run, but with the sequence of stimuli reversed: target plus maskers followed by target alone. This had very little effect on the pattern of the results. Also, subjects reported that the target frequency was still very easy to hear in all conditions. It appears that simply hearing the target on its own, either before or after the complex mixture, makes it easier to hear out the target from the complex. Of course, in the masking experiments the target was never heard alone immediately before or after presentation with a masker. Thus, subjects may have had trouble hearing out the target frequency in the masking experiments, but not in the rating experiments.

To check on this possibility, some of the conditions of experiment 1 were re-run, but with a 'cue' tone presented before each forced-choice trial. The cue tone had the same duration, level and frequency as the target, but it was unmodulated. The interval between the cue tone and the first test stimulus was 300 ms. Subjects were told that the cue tone had the same frequency as the target tone whose modulation depth they were required to discriminate. Subjects reported that the cue tone did make it easier to hear out the target frequency from the complex mixture. Furthermore, thresholds with the cue tone were lower than those obtained without the cue tone. This suggests that part of the MDI observed in experiments 1 and 3 may have been caused by difficulty in hearing out the frequency of the target tone. However, the decrease in MDI produced by the cue tone was relatively small. For example, for 230 and 3300 Hz maskers modulated in phase with the target at a 10 Hz rate, the mean value of delta *m* for a duration of 400 ms was 0.26 without the cue tone and 0.23 with the cue tone. The value of delta *m* without any masker was 0.051, so it is clear that substantial MDI occurred even with the cue tone. The only exception was with the unmodulated masker, where thresholds with the cue tone were only slightly higher than those obtained without any masker.

It appears, then, that MDI arises in two ways: (i) from difficulty in hearing a pitch corresponding to the frequency of the target; (ii) from difficulty in hearing the modulation of the target, or distinguishing that modulation from the modulation of other components present in the sound. We will denote the former 'carrier-specific' MDI and the latter 'modulation-specific' MDI. Carrier-specific MDI can occur both for unmodulated and for modulated maskers, but it is reduced or abolished by conditions designed to promote perceptual segregation of the target and maskers, such as those using onset and offset asynchronies or using a cue tone. Modulation-specific MDI occurs mainly for modulated maskers, and it persists under conditions designed to promote perceptual segregation of the target and maskers. It remains somewhat unclear whether carrier-specific MDI results partly from perceptual grouping caused by common or similar modulation; our results suggest that common gating is a more powerful factor.

Some indication of the possible nature of modulation-specific MDI is provided by an experiment of Hall & Grose (1991). Their subjects were presented with 1 and 2 kHz tones, gated synchronously. On each trial, one of the tones, selected at random, was amplitude modulated and the other was unmodulated. The task of the subjects was to identify which tone was modulated. Remarkably, most subjects performed rather poorly at this task. Even when the modulation was clearly audible, it was hard to say whether the higher or lower tone was modulated. It appears that subjects sometimes have difficulty assigning modulation to its appropriate carrier frequency.

In our experiments 2 and 4, subjects may have been able to listen for the overall clarity of 10 Hz modulation, but they may have had difficulty telling which carrier had that modulation. When all carriers were modulated at 10 Hz, the 10 Hz modulation would obviously have been easy to hear. When the maskers were modulated at 4 Hz or 25 Hz, the overall impression of hearing 10 Hz modulation definitely decreased, and clarity ratings were lower. When the maskers were unmodulated, this also reduced the overall impression of hearing 10 Hz modulation, again giving lower clarity ratings (in experiment 4).

The mechanisms underlying modulation-specific MDI remain unclear. One possibility is that it reflects the operation of 'channels' specialised for detecting and analysing modulation. There is both physiological evidence (Kay 1982; Rees & Moller 1983; Schreiner & Urbas 1986) and psychophysical evidence (Kay & Mathews 1972; Tansley & Suffield 1983; Bacon & Grantham 1989; Houtgast 1989) for such channels. Yost *et al.* (1989) suggested that MDI might arise in the following way. The stimulus is first processed by an array of auditory filters. The envelope at the output of each filter is extracted. When modulation is present, channels are excited that are tuned for modulation rate. All filters responding with the same modulation rate excite the same channel, regardless of the filter centre frequency. Thus, modulation at one centre frequency can adversely affect the detection and discrimination of modulation at other centre frequencies. Hall & Grose (1991) pointed out that such channels could provide an explanation for their finding that subjects have difficulty assigning modulation to its appropriate carrier frequency. They could also account for the general form of the rating data obtained with modulated maskers in experiment 2.

The function of the modulation channels remains unclear. Yost *et al.* (1989) suggested that they could provide a means for perceptual grouping of components with common modulation patterns. However, our data suggest that modulation-specific MDI persists under conditions where the target is perceptually segregated from the masker. In addition, the tuning of MDI in the modulation domain appears to be too broad for it to be useful for perceptual grouping.

It should be emphasized that such modulation channels can only explain part of the MDI observed in our experiments, namely modulation-specific MDI. For example, they do not account for the MDI produced by unmodulated maskers; such maskers should not excite the modulation channels. Also it is difficult to explain the effects of onset and offset asynchrony in terms of such channels. To do this it would be necessary to assume that the channels are not 'hard wired' but operate after perceptual grouping processes.

In summary, it appears that MDI may have more than one cause. One component of MDI, carrier-specific MDI, is caused by difficulty in hearing a pitch corresponding to the target frequency. Carrier-specific MDI appears to be strongly affected by perceptual grouping. Its effects can be reduced or abolished by asynchronous gating of the target and masker or by presenting a cue tone at the target frequency. The second component, modulation-specific MDI, is caused by a specific difficulty in hearing the modulation of the target, or in distinguishing that modulation from the modulation of other sounds that are present, or in assigning modulation to the appropriate carrier frequency. Modulation-specific MDI occurs even under conditions promoting perceptual segregation of the target and masker. It may reflect the operation of channels tuned for detecting modulation but very broadly tuned for carrier frequency.

This work was supported by the Medical Research Council (U.K.). We thank Brian Glasberg for writing the computer programs to run these experiments, and Tom Baer, Bob Carlyon, Joe Hall, Brian Glasberg and Michael Stone for helpful comments on an earlier version of this paper.

REFERENCES

Bacon, S.P. & Grantham, D. W. 1989 Modulation masking: effects of modulation frequency, depth and phase. *J. acoust. Soc. Am.* **85**, 2575–2580.

Bregman, A.S. 1990 *Auditory scene analysis: the perceptual organisation of sound.* Cambridge, Massachusetts: Bradford Books, MIT Press.

Bregman, A.S., Levitan, R. & Liao, C. 1990 Fusion of auditory components: effects of the frequency of amplitude modulation. *Percept. Psychophys.* **47**, 68–73.

Bregman, A.S. & Pinker, S. 1978 Auditory streaming and the building of timbre. *Can. J. Psychol.* **32**, 19–31.

Darwin, C.J. 1981 Perceptual grouping of speech components differing in fundamental frequency and onset time. *Q. Jl exp. Psychol.* **33A**, 185–287.

Darwin, C.J. & Sutherland, N.S. 1984 Grouping frequency components of vowels: when is a harmonic not a harmonic? *Q. Jl exp. Psychol.* **36A**, 193–208.

Fletcher, H. 1940 Auditory patterns. *Rev. Mod. Phys.* **12**, 47–65.

Hall, J.W. & Grose, J.H. 1991 Some effects of auditory grouping factors on modulation detection interference (MDI). *J. acoust. Soc. Am.* **90**, 3028–3035.

Houtgast, T. 1989 Frequency selectivity in amplitude-modulation detection. *J. acoust. Soc. Am.* **85**, 1676–1680.

Kay, R.H. 1982 Hearing of modulation in sounds. *Physiol. Rev.* **62**, 894–975.

Kay, R.H. & Mathews, D.R. 1972 On the existence in human auditory pathways of channels selectively tuned to the modulation present in frequency modulated tones. *J. Physiol., Lond.* **225**, 657–667.

McAdams S. 1984 Spectral fusion, spectral parsing and the

formation of the auditory image. Ph.D. thesis, University of Stanford.

Moore, B.C.J. 1989 *An introduction to the psychology of hearing*, 3rd edn. London: Academic Press.

Moore, B.C.J., Glasberg, B.R., Gaunt, T. & Child, T. 1991 Across-channel masking of changes in modulation depth for amplitude- and frequency-modulated signals. *Q. Jl exp. Psychol.* **43A**, 327–347.

Moore, B.C.J., Glasberg, B.R. & Peters, R.W. 1986 Thresholds for hearing mistuned partials as separate tones in harmonic complexes. *J. acoust. Soc. Am.* **80**, 479–483.

Moore, B.C.J., Glasberg, B.R. & Schooneveldt, G.P. 1990 Across-channel masking and comodulation masking release. *J. acoust. Soc. Am.* **87**, 1683–1694.

Moore, B.C.J. & Jorasz, U. 1992 Detection of changes in modulation depth of a target sound in the presence of other modulated sounds. *J. acoust. Soc. Am.* **91**, 1051–1061.

Patterson, R.D. & Moore, B.C.J. 1986 Auditory filters and excitation patterns as representations of frequency resolutions. In *Frequency selectivity in hearing* (ed. B. C. J. Moore), pp. 123–177. London: Academic Press.

Rasch, R.A. 1978 The perception of simultaneous notes such as in polyphonic music. *Acustica* **40**, 21–33.

Rees, A. & Moller, A.R. 1983 Responses of neurons in the inferior colliculus of the rat to AM and FM tones. *Hear. Res.* **10**, 301–310.

Roberts, B. & Moore, B.C.J. 1991 The influence of extraneous sounds on the perceptual estimation of first-formant frequency in vowels under conditions of asynchrony. *J. acoust. Soc. Am.* **89**, 2922–2932.

Roberts, B.R. & Bregman, A. 1991 Effects of the pattern of spectral spacing on the perceptual fusion of harmonics. *J. acoust. Soc. Am.* **90**, 3050–3060.

Schreiner, C.E. & Urbas, J.V. 1986 Representation of amplitude modulation in the auditory cortex of the cat. I. The anterior auditory field (AAF). *Hear. Res.* **21**, 227–241.

Tansley, B.W. & Suffield, J.B. 1983 Time course of adaptation and recovery of channels selectively sensitive to frequency and amplitude modulation. *J. acoust. Soc. Am.* **74**, 765–775.

Wilson, A.S., Hall, J.W. & Grose, J.H. 1990 Detection of frequency modulation (FM) in the presence of a second FM tone. *J. acoust. Soc. Am.* **88**, 1333–1338.

Yost, W.A. & Sheft, S. 1989 Across-critical-band processing of amplitude-modulated tones. *J. acoust. Soc. Am.* **85**, 848–857.

Yost, W.A., Sheft, S. & Opie, J. 1989 Modulation interference in detection and discrimination of amplitude modulation. *J. acoust. Soc. Am.* **86**, 2138–2147.

The psychophysics of concurrent sound segregation

ROBERT P. CARLYON

Laboratory of Experimental Psychology, University of Sussex, Brighton BN1 9QG, U.K.

SUMMARY

To perceptually separate concurrent complex sounds, normally hearing listeners simultaneously combine information across a wide range of frequency components. Three psychoacoustical experiments are described which investigate different forms of this across-frequency processing. The first two experiments investigate the role of coherence of frequency modulation (FM) between widely separated frequency components of a complex sound. The first experiment bolsters existing evidence that, for harmonic sounds, listeners can discriminate coherent from incoherent FM, but only by detecting the mistuning that arises from incoherent FM. The second demonstrates that, for inharmonic sounds, coherence of FM has no effect on the phenomenon of modulation detection interference (see Moore & Shailer, this symposium) once within-channel cues (combination tones and beating) are masked by background noise. It is concluded that there is not an across-frequency mechanism specific to the detection of FM incoherence. The third experiment investigates the extent to which the detection of mistuning of one component of a harmonic complex is impaired by an interfering sound (the 'interferer') with a frequency spectrum similar to that of the mistuned component. When the interferer is gated on and off with the harmonic complex, it has only a small effect provided that its level is more than 3 dB below that of the target. However, when the interferer starts before and ends after the complex, thresholds are elevated more, and this elevation occurs even for low-level interferers. Explanations of this effect in terms of adaptation and of auditory streaming are discussed.

1. INTRODUCTION

Until early last decade, most psychoacoustical research concentrated on the ability of listeners to make sequential comparisons between pairs of stimuli that differed along a single dimension, such as the intensity or frequency of a pure tone. Such experiments allowed the development of quite precise theories of basic auditory processes, and of models that could accurately predict experimental data. For example, the threshold for a pure tone signal in a masking band of noise can be accurately predicted from the amount of masker energy passing through an 'auditory filter' centered on the tone (Fletcher 1940; Zwicker *et al.* 1957; Patterson 1976). More recently, psychoacousticians have developed techniques for studying the processes used in many everyday listening situations, where we have to perform simultaneous comparisons of energy falling in different frequency regions. Across-frequency processing is important for at least two real-life tasks. First, to identify the spectral shape of an isolated sound (e.g. a vowel), one has to compare the levels of different frequency components. Perhaps more importantly, the perceptual separation of two sounds with overlapping spectra requires listeners to 'sort through' the combined spectrum, identifying which frequency components belong to which sound. The ability of listeners to perform both of these types of simultaneous, across-frequency comparisons, has been the subject of much recent experi-

mental work (Hall *et al.* 1984; Moore & Glasberg 1986; Demany & Semal 1988; Green 1988; Carlyon & Stubbs 1989; Demany *et al.* 1991). The experiments reported here investigate two potential cues to concurrent sound segregation that require across-frequency processing.

(a) F_0 differences

Periodic sounds, such as the vowels of speech contain frequency components ('harmonics') equal to integer multiples of a common fundamental ('F_0'). Consequently, frequency components that do not correspond to one of these harmonics can be attributed to a different sound source, and can be discriminated from an in-tune harmonic, even when all harmonics excite separate auditory filters (Moore *et al.* 1985b; Moore & Glasberg 1986; Demany & Semal, 1988, 1991; Hartmann *et al.* 1990; Demany *et al.* 1991; Carlyon *et al.* 1992). It is also known that listeners can use F_0 differences between competing sources of voiced speech to improve the identification of the constituent sounds (Brokx & Nooteboom 1982; Scheffers 1983; Summerfield & Assmann 1991).

(b) FM coherence

A second, attractive, potential cue for the perceptual grouping of different components of a complex sound is FM coherence. This refers to the fact that

Phil. Trans. R. Soc. Lond. B (1992) **336**, 347–355
Printed in Great Britain

when the fundamental frequency (F_0) of, for example, a speaker's voice changes, all the frequency components of that voice change in the same direction at the same time. It is plausible that listeners group together the coherently changing components of a single voice, and separate them from components arising from a different source, which may be modulated incoherently with that voice. If so, the auditory system might use FM coherence in the same way that the visual system processes coherence of spatial movement to group different parts of the same object. By analogy with vision, we might imagine a mechanism whereby the auditory system not only identifies the peaks in the basilar membrane excitation pattern caused by individual components, but also correlates their movement along it (McAdams 1984; Bregman 1990; Wilson *et al.* 1990). However, it is difficult to show conclusively that FM incoherence per se can facilitate the perceptual separation of frequency components because, when a component of a harmonic complex is modulated incoherently from the others, it also becomes mistuned. Listeners might detect this mistuning, rather than the FM coherence per se.

The first experiment reported here extends previous work suggesting that, once within-channel cues are removed, listeners are not sensitive to FM incoherence, independently of the mistuning that it causes. The second experiment resolves a discrepancy between the results of experiment 1 and those of a study (Wilson *et al.* 1990) that investigated the influence of FM incoherence on the phenomenon of 'modulation detection interference' (see Moore & Shailer, this symposium). Finally, a third experiment measures thresholds for the detection of mistuning imposed on one component of a harmonic complex, both in the presence and absence of an interfering sound. The 'interferer' has a frequency spectrum similar to that of the potentially mistuned harmonic, and is either gated synchronously with the complex or starts before and ends after it.

2. DETECTION OF FM INCOHERENCE

(a) Background

In a recent article (Carlyon 1991), it was argued that listeners are not sensitive to FM coherence in the absence of additional cues such as mistuning. In one experiment, it was shown that although listeners could discriminate between coherent and incoherent sinusoidal FM of harmonic complexes, they could not do so when the complexes were inharmonic. A second experiment showed that listeners' psychometric functions for the detection of a static mistuning imposed on one component of a harmonic complex could account for the corresponding functions describing the detection of FM coherence. The article concluded that there is no across-frequency mechanism specific to the detection of FM incoherence.

The finding that listeners cannot discriminate coherent from incoherent FM, if generalizable to all stimuli, demonstrates that across-frequency comparisons of FM coherence cannot play a role in perceptual sound segregation. However, it is possible that the

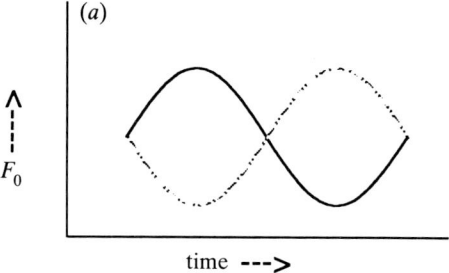

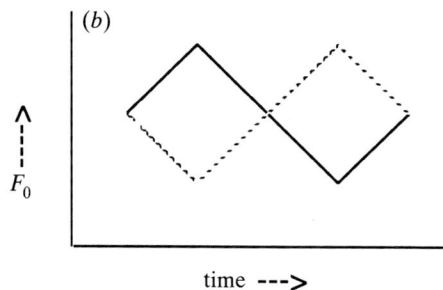

Figure 1. Schematic representation of the pattern of F_0 change imposed by two patterns of FM. (*a*) Sinusoidal FM; (*b*) triangular FM. The two traces in each part of the figure represent the F_0s of two sounds that are modulated incoherently (π modulator delay).

findings reported by Carlyon (1991) were specific to the sinusoidal nature of the FM used. As figure 1*a* shows, imposing incoherent sinusoidal FM on two sounds causes their F_0s to spend a moderate amount of time at their maximum mistuning. During this time the F_0s are roughly constant, and therefore do not change incoherently: thus, the choice of a sinusoidal pattern of modulation could have biased listeners away from detecting incoherence, and towards the detection of mistuning. The aim of the first experiment reported here was to test this explanation by comparing the detection of incoherence obtained with sinusoidal modulation with that obtained using triangular FM. As figure 1*b* shows, incoherent triangular FM results in the two F_0s reaching their maximum mistuning only momentarily, while spending virtually all their time moving in opposite directions. If listeners are sensitive to FM incoherence, then performance should be better with triangular than with sinusoidal FM. If, however, they are sensitive only to mistuning, then the opposite should be true. In addition, psychometric functions were measured for the detection of static mistunings, using the two types of modulation pattern. If listeners detect incoherence only from the mistuning that it causes, then, for a given modulation pattern, psychometric functions for the detection of incoherence and of static mistunings should be similar.

(b) Stimuli and procedure

The method of stimulus generation and procedure were as described by Carlyon (1991), except for the inclusion of a condition with triangular modulation. Briefly, a three-interval, two-alternative forced-choice

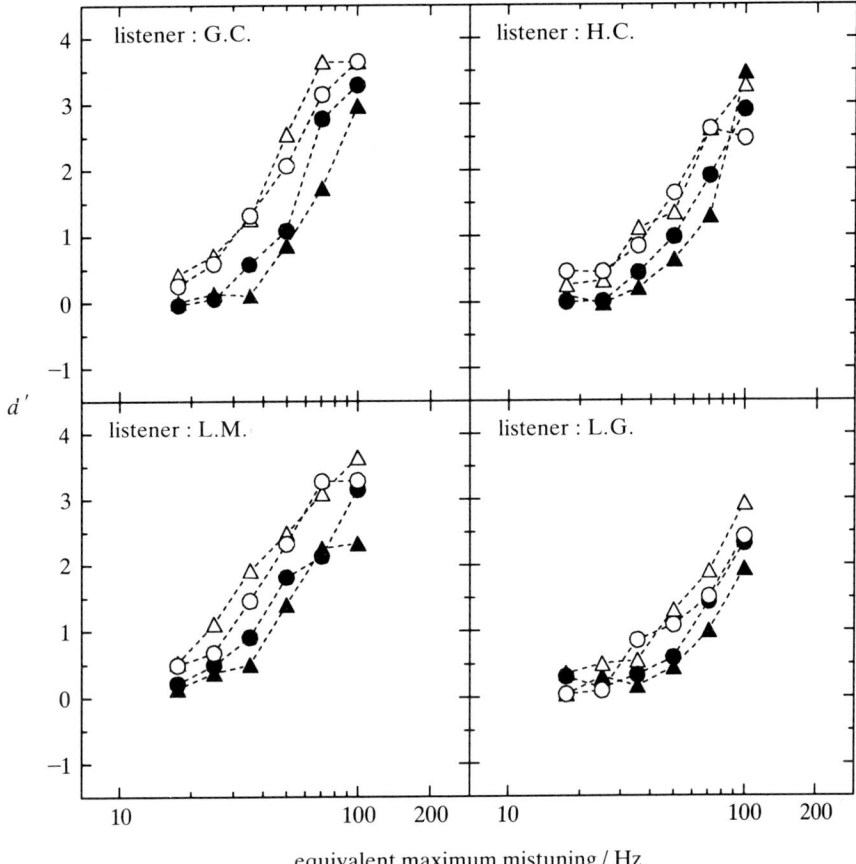

Figure 2. Sensitivity (d') as a function of modulator phase delay for each of four listeners. Data obtained from sinusoidal FM is shown by circles, that from triangular FM by triangles. Filled symbols are for the incoherence condition, and open symbols are for the mistuning condition.

paradigm with feedback was used. Stimuli consisted of three sinusoidal components (1500, 2000 and 2500 Hz), each modulated by a 2.5 Hz sinusoidal or triangular wave, with a zero-peak modulation depth of 2.5% of the carrier frequency. Signal duration was 400 ms, and all stimuli were presented at a level of 45 dB SPL per component in a background of pink noise (spectrum level 12.2 dB SPL at 2000 Hz).

The standard (non-signal) stimulus always consisted of the three sinusoidal components, in tune and modulated coherently. In the 'mistuning' conditions, sensitivity (d') was measured as a function of a static mistuning (± 18, 25, 35, 50, 71, 100 Hz) imposed on the 2000 Hz component in the signal interval. In the 'incoherence' conditions, d' was measured as a function of the modulator phase delay (0.11 to 1.0 π radians) imposed on the 2000 Hz component. The modulator delays were chosen so that the maximum mistunings that they produced (at any time during the signal) were equivalent to the static mistunings imposed in the other condition (Carlyon 1991); accordingly, data in both conditions will be discussed and plotted in terms of 'equivalent maximum mistunings (EMMS)'. Plotting data in this way allows us to make a specific prediction: if listeners detect FM incoherence from the mistuning that it causes, then the psychometric functions in the mistuning and incoherence conditions should be parallel, as we

would expect sensitivity to vary with EMM in the same way for both conditions. In all conditions, the playback rate, and hence the nominal F_0, was randomized by $\pm 5\%$ from presentation to presentation, and the overall modulator starting phase was randomized from zero to π radians.

(c) *Results*

Sensitivity (d') is shown as a function of EMM for each of the four listeners in figure 2. One important comparison is in the incoherence condition, between performance with sinusoidal (filled circles) and with triangular (filled triangles) modulation. Performance for all four listeners is consistently better with sinusoidal modulation, indicating that they were indeed detecting mistuning, rather than FM incoherence per se (planned comparisons†, $F = 7.449$, $p < 0.05$). This conclusion is supported by the observation that for both types of modulation, the slopes of the psychometric

† The planned comparisons were performed after a two-way ANOVA (four 'modulation conditions' (combinations of modulator type and incoherence or mistuning) × six 'equivalent maximum mistunings (EMMS)'). The ANOVA revealed significant main effects of modulation condition ($F = 22.055$, $p < 0.001$), of EMM ($F = 64.042$, $p < 0.001$). The interaction between modulation condition and EMM was not significant.

functions in the incoherence (filled symbols) and mistuning (open symbols) conditions do not differ significantly (*t*-tests, c.f. Edwards 1973). Both of these results support the conclusions drawn in the Carlyon (1991) paper that, for harmonic stimuli, listeners detect FM incoherence from the mistuning that it causes. Taken together with the finding that listeners cannot detect FM incoherence imposed on a component of an inharmonic complex, this provides strong evidence against an across-frequency mechanism specific to the detection of FM incoherence.

3. MODULATION DETECTION INTERFERENCE

Elsewhere in this publication, Moore & Shailer describe a phenomenon, termed modulation detection interference (MDI), which reflects the across-frequency processing of complex sounds. In a typical MDI experiment, listeners are required to detect amplitude or frequency modulation of a sinusoidal carrier in the presence of an interfering sinusoid, whose frequency is such that it does not mask the signal. A common finding is that although the 'interferer' has no effect when it is unmodulated, a modulated interferer increases the signal's modulation detection threshold (Yost & Sheft 1989; Yost *et al.* 1989; Wilson *et al.* 1990; Moore *et al.* 1991). Thus, although the interferer has no effect on the signal in a detection experiment, it affects performance in a supra-threshold task such as the detection of modulation. In a recent paper, Wilson *et al.* (1990) reported that, for the detection of FM, threshold was affected not only by the imposition of FM on the interferer, but also by the amount of incoherence (modulator phase delay) between the signal and interferer modulation: coherent FM raised thresholds more than did incoherent FM. They suggested that, with coherent FM, listeners grouped the interferer with the signal, and that this impaired the detection of FM. With incoherent FM, the interferer would not be so strongly grouped with the signal, and so the FM detection threshold would be elevated less.

The dependence of MDI on FM coherence seems, at least at first sight, to contradict the conclusions drawn in the first part of this article: if listeners are not sensitive to FM coherence, how can it affect FM detection thresholds? Wilson *et al.*'s stimuli were inharmonic, with signal and interferer frequencies of 1900 Hz and 2500 Hz respectively, so it is unlikely that their results were due to differences in harmonicity between coherently and incoherently modulated sounds. However, there are two other possible reasons why they may have arrived at a different conclusion from that drawn here. Both arise from the observation that, whereas we presented our stimuli at a level of 45 dB SPL per component in a pink noise background, Wilson *et al.*'s sounds were presented at 65 dB SPL per component in quiet.

The first possibility was suggested by a pilot experiment in which I tried to discriminate between coherently and incoherently modulated versions of Wilson *et al.*'s stimuli. I found that I could do so, but only by listening for a prominent combination tone

(CT) in the incoherent case: when a 1400 Hz lowpass noise was added to mask the CTs, my performance dropped to chance. There are at least two ways in which CTs could affect the amount of MDI. First, in the incoherent condition, a CT with a frequency such as $f_2 - f_1$ or $2f_1 - f_2$ would be modulated over a much wider frequency range in the signal than in the non-signal interval, and listeners could detect this prominent-sounding CT, thereby lowering thresholds and hence the amount of MDI. Second, in the standard (non-signal) interval of a two-interval, forced-choice trial, the combination tone $f_2 - f_1$ will be frequency modulated over the same range as the interferer. Moore *et al.* (1991) have reported that the existence of a second interfering tone increases the amount of MDI, and it is possible that the (modulated) CT raised modulation detection thresholds in a similar manner to an externally presented tone. As the stimuli in the standard interval are very similar in the coherent and incoherent conditions, this additional effect would be the same in the two conditions. Thus there are two possible effects of CTs; one which might reduce MDI in the incoherent condition, and a second which would tend to raise MDI in both conditions.

An alternative explanation for Wilson *et al.*'s finding arises from beating between the target and interfering tones in auditory filters that respond to both components. The beating is most likely to occur in filters with centre frequencies (CFs) between the target and interferer frequencies. In one of their experiments, Wilson *et al.* (1990) interspersed a narrowband noise between the target and interferer tones, and showed that the MDI caused by coherent FM persisted, and was therefore not mediated by within-channel beating. However, with incoherent FM, the within-channel beating will be stronger than in the coherent condition, and stronger than when the target is not modulated (i.e. in the non-signal interval). This is because during incoherent FM the component frequencies first move away from each other, causing the output of a filter tuned between them to decrease, and then move towards each other, causing the filter output to increase. Listeners might use this 'within-channel amplitude modulation' to detect the signal in the incoherent condition, thereby reducing their MDI. Thus, even though beating might not be necessary for the basic MDI phenomenon, it might account for the difference in MDI obtained with coherent and incoherent FM.

Experiment 2 investigated whether Wilson *et al.*'s finding of greater MDI for coherent than for incoherent FM was really due to an across-frequency grouping mechanism as they suggest, or whether it can be attributed to the detectability of CTs and within-channel beating.

(b) Method

The first set of conditions was a direct replication of Wilson *et al.*'s experiment 3. Using an adaptive procedure (Levitt, 1971), the threshold FM depth necessary for listeners to discriminate between a 1900 Hz carrier that was either unmodulated, or sinusoidally frequency-modulated at a rate of 6 Hz, was measured in the presence of an interfering 2500 Hz tone. The interferer was presented in both intervals of each 2I, 2AFC trial, and was either unmodulated, modulated coherently with the signal (6 Hz, zero-peak FM depth fixed at 2.5% of carrier frequency), or modulated incoherently with the signal. Incoherent FM was produced by introducing a delay of π radians between the interferer and signal modulators. Both the interferer and the target component were presented at a level of 65 dB SPL in quiet. In a second set of conditions, CTS were masked by presenting the stimuli in a background of continuous lowpass noise, generated by passing a pink noise through two lowpass filters (Kemo VB25.03; attenuation 48 dB per octave each) in series. The noise had a spectrum level of 32.2 dB SPL at 1 kHz, and the filter cutoffs (3-dB-down) were set to 1400 Hz. In a third set of conditions, both CTS and within-channel interactions were masked by a 5-kHz-wide pink noise, whose spectrum level at 1 kHz was also 32.2 dB SPL.

(c) Results

The data for stimuli presented in quiet are shown in figure 3a. The existence of MDI is confirmed by the observation that, for four out of five listeners, thresholds are greater with a coherently modulated interferer (open bars), than when the interferer is unmodulated (filled bars). Also, the general pattern of results reported by Wilson *et al.* is confirmed, in that thresholds with incoherent FM are generally lower than with coherent FM (compare shaded with open bars). The contribution of CTS to this effect of FM coherence is shown in figure 3b, which shows that, when stimuli are presented in lowpass noise, the effect persists only for two out of the five listeners (G.C. and L.G.). Note that CTS do not seem to be essential for the basic MDI effect: thresholds obtained with an unmodulated interferer are still lower than in the presence of a coherently modulated interferer, even in the presence of lowpass noise. Finally, figure 3c shows that when within-channel interactions are also eliminated, by adding wideband pink noise instead of lowpass noise, all five listeners show essentially identical thresholds with coherent and incoherent interferers. For four listeners, thresholds with both of these modulated interferers are higher than in the unmodulated condition. The results of experiment 2 confirm that CTS and within-channel cues are not essential for the basic MDI phenomenon (Yost & Sheft 1989; Yost *et al.* 1989; Wilson *et al.* 1990; Moore *et al.* 1991), but that they are responsible for the dependence of MDI on FM coherence.

4. DISCRIMINATING BETWEEN TUNED AND MISTUNED HARMONICS IN THE PRESENCE OF A COMPETING SOUND

(a) Motivation and general description

The experiment described in the first part of this article added to the existing evidence that listeners can

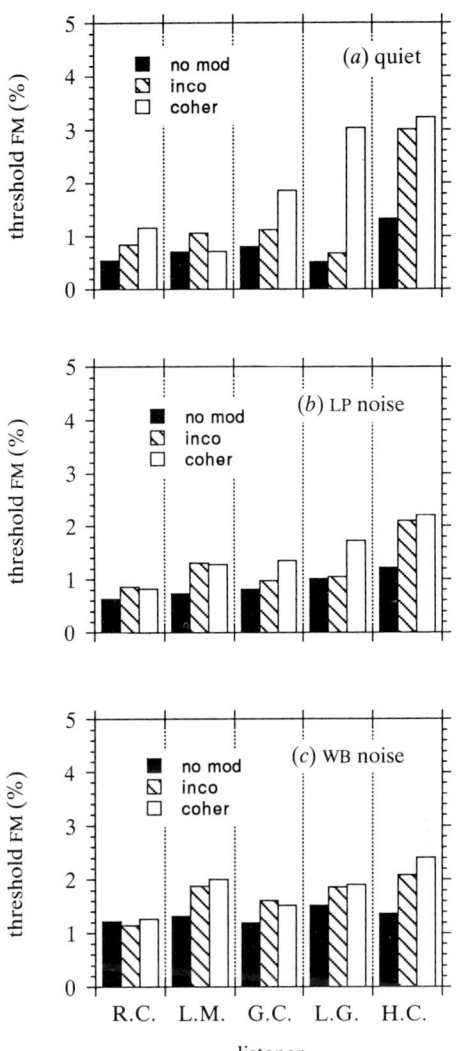

Figure 3. Each panel shows modulation detection thresholds, expressed as a percentage of the carrier frequency, for five listeners (spaced along the abscissa). For each listener, three thresholds are shown, corresponding to the modulation imposed on the interferer. Filled bars represent no modulation, open bars coherent modulation, and circles incoherent modulation. Modulation detection interference for the two conditions with modulated interferers is equal to the threshold elevation re. the condition with an unmodulated interferer. (a) Stimuli presented in quiet; (b) 1400 Hz lowpass noise background; (c) wideband pink noise background.

detect a mistuning imposed on one component of an otherwise harmonic complex tone (Moore *et al.* 1985b; Moore & Glasberg, 1986; Demany *et al.* 1991; Carlyon *et al.* 1992). The fact that we can 'hear out' a mistuned harmonic that is resolved from its neighbours shows that the auditory system can stimultaneously compare the frequencies of resolved harmonics, and can use this information to perceptually separate one component of a complex sound from the rest. Both Moore *et al.* (1985b) and Demany *et al.* (1991) have shown that listeners can detect mistunings of individual resolved components of as little as 1%; however, these measurements were made either in quiet (Moore *et al.* 1985b), or in the presence of a low-level pink-noise background (Demany *et al.* 1991). In real life, even when a component of a harmonic complex is resolved from its neighbours, there will sometimes exist a component of a different complex (e.g. a competing vowel sound) with a frequency very close to that of the target. One aim of our third experiment was to determine the extent to which detection of mistuning is affected by an interfering sound. We also investigated whether sensitivity to mistuning was affected by the existence of onset and offset asynchronies between the interfering sound and the harmonic complex.

(b) Method

Experiment 3 took advantage of the fact that one can measure sensitivity to mistuning by frequency modulating one component of a harmonic complex incoherently from the other components (cf. experiment 1). The harmonic complex consisted of the first seven harmonics of 500 Hz, presented at the same level and in the same background noise as the stimuli in experiment 1, and with a duration of 200 ms. In the standard interval of each 21, 2AFC trial, all components were frequency modulated coherently at a rate of 5 Hz; in the signal interval, the 'target component' (the fourth harmonic, frequency = 2000 Hz) was modulated incoherently (π radians modulator delay). The amount of mistuning is proportional to the FM depth imposed on all components, so the threshold FM depth was determined using an adaptive procedure (Levitt 1971).

In some conditions, there was an interfering sinusoid present in both intervals of each trial; its level was -9 dB, -6 dB, -3 dB, or 0 dB relative to that of the components of the harmonic complex. The 'interferer' was either turned on and off at the same time as the complex (condition 'SYNCH'; figure 4*a*), or was turned on 400 ms before its onset and turned off 100 ms after its offset (condition 'ASYNCH'; figure 4*b*). It should be obvious from figure 4*a*, *b*, that within-channel interactions (such as beating) will occur between the interferer and the target (2000 Hz) component, and that their patterning will depend on the phase of the target modulation. It is necessary to ensure that listeners do not identify the signal interval using such a within-channel cue. Therefore, the overall starting modulation phase of the stimuli was randomized from presentation to presentation: this is illustrated in figure 4*c*, which shows a different possible trial structure in the 'SYNCH' condition. Note that any particular pattern of interaction between target and interferer is equally likely to occur in the signal and standard intervals.

In an additional condition, the interfering sinusoid was replaced by a 200-Hz-wide noise, with an overall level 6 dB below that of the target component. It was gated on either synchronously or asynchronously with the target.

(c) Results

Figure 5 shows the threshold FM depth for the SYNCH (open triangles) and ASYNCH (open squares) condi-

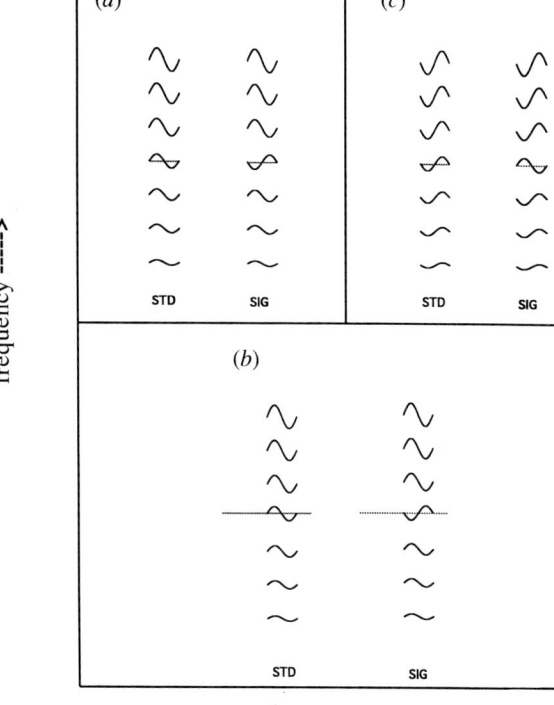

Figure 4. Panels (*a*) and (*c*) show schematic spectrograms of two possible trial structures in the SYNCH condition of experiment 3. Panel (*b*) shows one of the possible trial structures in the ASYNCH condition.

tions, as a function of the level of the interferer relative to that of the target component. In the SYNCH condition, thresholds are not substantially affected by interferers with levels re. the target of -9 or -6 dB, but rise steadily with further increases in interferer level. Perhaps the most striking finding is that, in the ASYNCH condition, thresholds are higher than in the SYNCH condition, and are affected by interferers with levels as low as -9 dB re. the target. This is true not only for an interfering sinusoid, but also for an interfering narrowband noise (compare the filled triangles and squares). The reasons for this additional effect of interferer asynchrony (over and above that caused by a synchronous interferer) will be discussed in § 5*b*.

5. DISCUSSION

(a) Why can't listeners discriminate coherent from incoherent FM?

Experiments 1 and 2 supported the conclusion reached by Carlyon (1991) that there is not an across-frequency mechanism specific to the detection of FM incoherence. Given that FM incoherence seems such an attractive cue for segregation, the question arises as to why listeners cannot detect it in the absence of other cues. It seems likely that this is due to a number of reasons.

First, Summerfield (1991, see also this symposium) has pointed out that a mechanism specific to the detection of FM incoherence would be computationally

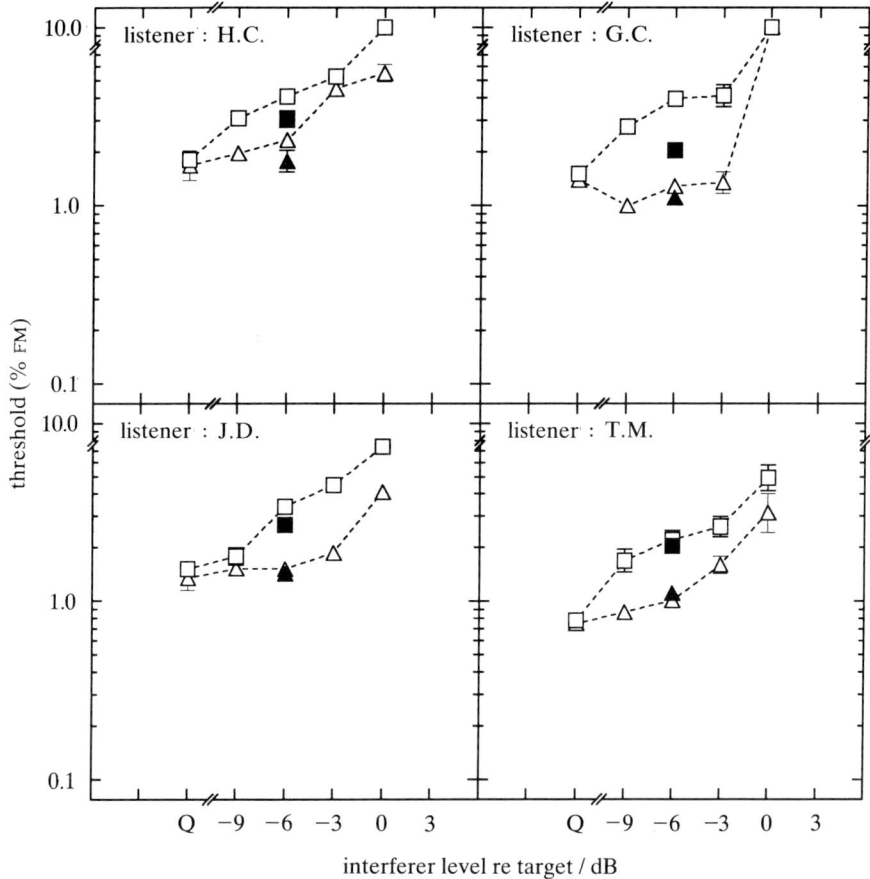

Figure 5. Modulation depth (as a percentage of carrier frequency) necessary for listeners to detect mistuning imposed on the fourth harmonic of a seven component complex ($F_0 = 500$ Hz). Each panel shows the data of one listener. Triangles, SYNCH condition; squares, ASYNCH condition. The open symbols are for a sinusoidal interferer, and the unconnected filled symbols are for a narrowband noise interferer.

expensive. He notes that most sounds that one would want to segregate on this basis are harmonic, and that as FM incoherence always leads to inharmonicity (for which there already exists a mechanism), it would be ecologically disadvantageous to develop a separate mechanism that provides no new information.

Second, it is worth noting recent evidence that coherent FM enhances grouping between a complex tone and an extra component, relative to the case with no FM. C. J. Darwin and V. Ciocca (personal communication) repeated an experiment by Moore *et al.* (1985), in which listeners were asked to adjust the pitch of an (unmodulated) harmonic complex to that of another complex which had had one component mistuned. They replicated Moore *et al.*'s finding for steady tones that, as the component was mistuned by up to about 3%, the pitch of the complex changed by increasing amounts in the same direction, whereas for larger mistunings the mistuned harmonic became segregated from the complex, and had a progressively smaller effect on its pitch. Darwin and Ciocca reported that when the whole stimulus was frequency modulated, the mistuned harmonic could contribute to the pitch of the complex at greater mistunings than was the case with no FM. They concluded that listeners have a tendency to group together components that are modulated. If modulated components are grouped together, then the effects of this grouping will be

apparent only for coherent FM, as incoherent FM will lead to mistuning, which will in turn counteract any 'grouping by modulation'. Thus, by a combined sensitivity to mistuning and to modulation, listeners could fuse coherently modulated harmonic components, and separate incoherently modulated components. This strategy would be computationally more efficient than correlating the movements of peaks in the excitation pattern.

Finally, it is worth noting that a 'peak correlation' mechanism for detecting FM incoherence could produce the 'wrong answer' when the peaks reflect maxima in the spectral envelope (e.g. formants), rather than in the spectral fine structure (e.g. harmonics). Unlike harmonics, formants of the same speech sound often change frequency in opposite directions ('incoherently'), even when they come from the same source. Thus, in order to work effectively, a correlation mechanism would have to be able to take into account whether a given peak was due to a harmonic or to a formant.

In summary, it is proposed that the auditory system groups components by harmonicity, supplemented by the presence, but not by the coherence, of FM. Such a mechanism makes ecological sense: it is rare, but possible, for steady components from different sources to be harmonically related, but such a co-incidence is much less likely when the components are frequency modulated. Furthermore, unlike a 'peak-correlation' mechanism, the proposed processing scheme would be both computationally efficient and unlikely to erroneously separate groups of components that come from the same source.

(b) The effect of interfering sounds on the detection of mistuning

Experiment 3 showed that a sinusoid that starts 400 ms before, and ends 100 ms after, a harmonic complex impairs the detection of mistuning more than does a synchronous interferer. This finding seemed puzzling at first: one might expect listeners to use the leading portion of the interferer to 'subtract' it from the percept of the complex sound, thereby improving their detection of mistuning. At the very least, one might expect trained listeners such as ours to learn to ignore the leading and lagging portions of the tone. It is therefore worth considering reasons why they could not do this.

One possibility is that the leading portion of the tone induced adaptation in the frequency region of the target component, thereby reducing the level of its representation in the auditory periphery: it is known that frequency DLs increase at very low levels. However, figure 5 shows that the effect of onset asynchrony (ratio between ASYNCH and SYNCH thresholds) is as high at interferer levels of -9 dB as it is at -3 dB or 0 dB. If the leading portion of the interferer reduced the auditory-nerve response to the target component via adaptation, then we would expect the effect to be greatest at high interferer levels.

A second, more promising possibility is that the leading portion of the interferer captured the target component in the same perceptual 'stream' (see Bregman 1990), and that this impaired listeners' ability to compare the target frequency to that of the other components. A similar process might account for a recent finding reported by Green & Dai (1991), involving a different type of across-frequency comparison. They used a 'profile analysis' task, which requires listeners to detect an increment in the level of one component of a complex sound relative to that of the other components. They reported that threshold increased when the target component was turned on before, and off after, the rest of the complex compared to a condition with synchronous onsets and offsets. Further evidence for the central origin of the present effect comes from additional experiments in which the interfering sound was presented diotically, while the complex and target tones were presented monotically. These showed that mistuning detection was much worse than when all sounds were presented monotically. This may have been due to components with frequencies close to that of the interferer being pulled even more strongly into a separate perceptual stream.

It is notable that our listeners showed a synchrony effect even though the interferer and target 'sounded different' from each other when presented in isolation (one was a steady tone, the other frequency modulated), and that the effect persisted when this quality difference was increased by replacing the interferer

with a narrowband noise. A possible interpretation is that a stimulus with a similar excitation pattern as the interferer is captured into its perceptual stream, irrespective of its other properties (tone versus noise, steady versus FM). If so, then in our experiment the auditory system acted in a fairly unsophisticated way when forming the continuous energy around 2000 Hz into a single perceptual stream. This outcome might be related to the results of Roberts & Moore (1990), who showed that adding pairs of tones with frequencies slightly above the first formant of a vowel increased the perceived first formant frequency (all sounds were gated on and off together). They reported that a similar effect could be obtained by replacing the tone pairs with a narrowband noise, despite its 'radically different temporal structure and timbre' from the vowel sound.

In conclusion, experiment 3 has shown that sensitivity to mistuning persists not only when the harmonics are resolved by the peripheral auditory system, but also when a synchronous narrowband sound is present whose spectrum overlaps that of the mistuned component. However, this 'resilience' to the interferer can be impaired by other mechanisms when the interferer starts before, and ends after, the harmonic complex. The exact nature of the additional mechanism is yet to be determined, but we know that it acts centrally and is fairly insensitive to temporal properties of the interfering sound (e.g. tone versus noise). Further investigations, particularly into the reasons why even highly-trained listeners cannot 'ignore' the leading and lagging portions of the interferer, are currently underway.

The work reported here was supported by a Royal Society University Research Fellowship to the author. I thank Valter Ciocca and Quentin Summerfield for useful comments on an earlier version of this article. In particular, Valter Ciocca provided useful insights into the interpretation of experiment 3.

REFERENCES

Bregman, A.S. 1990 *Auditory scene analysis*. Cambridge, Massachusetts: M.I.T. Press.

Bregman, A.S., Levitan, R. & Liao, C. 1990 Fusion of auditory components: effects of the frequency of amplitude modulation. *Percept. Psychophys.* **47**, 68–73.

Brokx, J.P.L. & Nooteboom, S.G. 1982 Intonation and the perceptual separation of simultaneous voices. *J. Phonet.* **10**, 23–36.

Carlyon, R.P. 1991 Discriminating between coherent and incoherent frequency modulation of complex tones. *J. acoust. Soc. Am.* **89**, 329–340.

Carlyon, R.P., Demany, L. & Semal, C. 1992 Detection of across-frequency differences in fundamental frequency. *J. acoust. Soc. Am.* (In the press.)

Carlyon, R.P. & Stubbs, R.J. 1989 Detecting single-cycle frequency modulation imposed on sinusoidal, harmonic, and inharmonic carriers. *J. acoust. Soc. Am.* **85**, 2563–2574.

Demany, L. & Semal, C. 1988 Dichotic fusion of two tones one octave apart: Evidence for internal octave templates. *J. acoust. Soc. Am.* **83**, 687–695.

Demany, L. & Semal, C. 1991 Harmonic and melodic octave templates. *J. acoust. Soc. Am.* **88**, 2126–2135.

Demany, L., Semal, C. & Carlyon, R.P. 1991 Perceptual

limits of octave harmony and their origin. *J. acoust. Soc. Am.* **90**, 3019–3027.

Edwards, A.E. 1973 *Statistical methods.* New York: Holt, Rhinehardt, Winston.

Fletcher, H. 1940 Auditory patterns. *Rev. mod. Phys.* **12**, 47–65.

Green, D.M. 1988 *Profile analysis.* New York: Oxford University Press.

Green, D.M. & Dai, H. 1992 Temporal relations in profile comparisons. In *Auditory physiology and perception* (ed. Y. Cazals, L. Demany & K. Horner), pp. 471–478. Oxford: Pergamon Press.

Hall, J.W., Haggard, M.P. & Fernandes, M.A. 1984 Detection in noise by spectro-temporal pattern analysis. *J. acoust. Soc. Am.* **76**, 50–56.

Hartmann, W.M., McAdams, S. & Smith, B.K. 1990 Hearing a mistuned harmonic in an otherwise periodic complex tone. *J. acoust. Soc. Am.* **88**, 1712–1724.

Levitt, H. 1971 Transformed up-down methods in psychophysics. *J. acoust. Soc. Am.* **49**, 467–447.

McAdams, S. 1984 Spectral fusion, spectral parsing and the formation of the auditory image. Ph.D. thesis, University of Stanford.

Moore, B.C.J. & Glasberg, B.R. 1986 Thresholds for hearing mistuned partials as separate tones in harmonic complexes. *J. acoust. Soc. Am.* **80**, 479–483.

Moore, B.C.J., Glasberg, B.R., Gaunt, T. & Child, T. 1991 Across-channel masking of changes in modulation depth for amplitude- and frequency-modulated signals. *Q. J. exp. Psychol.* **43A**, 327–348.

Moore, B.C.J., Glasberg, B.R. & Peters, R.W. 1985 Relative dominance of individual partials in determining the pitch of complex tones. *J. acoust. Soc. Am.* **77**, 1853–1860.

Moore, B.C.J., Peters, R.W. & Glasberg, B.R. 1985*b* Thresholds for the detection of inharmonicity in complex tones. *J. acoust. Soc. Am.* **77**, 1861–1867.

Patterson, R.D. 1976 Auditory filter shapes derived with noise stimuli. *J. acoust. Soc. Am.* **59**, 640–654.

Roberts, B. & Moore, B.C.J. 1990 The influence of extraneous sounds on the perceptual estimation of first-formant frequency in vowels. *J. acoust. Soc. Am.* **88**, 2571–2583.

Scheffers, M.T.M. 1983 Sifting vowels: auditory pitch analysis and sound segregation. Ph.D. thesis, University of Groningen.

Summerfield, A.Q. 1991 Perception of concurrent vowels: effects of coherent frequency modulation. *Brit. J. Audiol.* (In the press.)

Summerfield, A.Q. & Assmann, P.F. 1991 Perception of concurrent vowels: Effects of pitch-pulse asynchrony and harmonic misalignement. *J. acoust. Soc. Am.* **89**, 1364–1377.

Wilson, A.S., Hall, J.W. & Grose, J.H. 1990 Detection of frequency modulation (FM) in the presence of a second FM tone. *J. acoust. Soc. Am.* **88**, 1333–1338.

Yost, W.A. & Sheft, S. 1989 Across-critical-band processing of amplitude-modulated tones. *J. acoust. Soc. Am.* **85**, 848–857.

Yost, W.A., Sheft, S. & Opie, J. 1989 Modulation interference in detection and discrimination of amplitude modulation. *J. acoust. Soc. Amer.* **86**, 2138–2147.

Zwicker, E., Flottorp, G. & Stevens, S.S. 1957 Critical bandwidths in loudness summation. *J. acoust. Soc. Am.* **29**, 548–557.

Discussion

S. McAdams (*Laboratoire de Psychologie Expérimentale, Université René Descartes, Paris, France*). The papers presented by Dr Summerfield and Dr Carlyon have presented rather convincing arguments against the direct involvement of coherent frequency modulation (cfm) in the perceptual segregation of concurrent sounds, the intimation being that FM does not account for any segregation over and above that which can be accounted for by the mistuning that such modulation would produce. If we accept this reasoning, however, there remains to be explained some puzzling experimental data. McAdams (1989) presented listeners with complex stimuli consisting of three vowels (/a/,/o/,/i/), each separated by five semitones (30%) which is well beyond the separation at which mistuning produces its maximum effect. Listeners were asked to judge the relative prominence of the vowels heard in the complex. Mean prominence judgments increased significantly when the vowels were modulated compared to when they were not. This finding suggests that something associated with FM made the vowels easier to hear in a complex background when modulated, in spite of the large static mistuning between them. The degree of increase in perceived prominence was the same when the vowels were: (i) modulated alone against a background of unmodulated vowels; (ii) modulated independently of the other vowels; or (iii) modulated coherently with the other vowels. Thus, whatever segregation was achieved by modulating a given vowel was no further enhanced by modulating the other concurrent vowels independently. Dr Carlyon's dichotomy between cues for grouping and cues for segregation claims that FM plays only a small role in grouping and none in segregation. According to this scheme then, the increased prominence would be due to the vowels' harmonics being perceptually grouped by FM. How, though, are we to explain the fact that grouping alone increases a vowel's perceived prominence in a complex mixture?

Reference

McAdams, S. 1989 Psychological constraints on form-bearing dimensions in music. *Contemp. Music Rev.* **4**, 181–198.

R. P. Carlyon. The finding Dr McAdams describes is of course consistent with those of Quentin Summerfield and myself, and represents some of the earliest and most convincing evidence that, although listeners are sensitive to the existence of FM, they are not sensitive to FM coherence between different concurrent sources.

It seems that, when a bunch of components are modulated by about the same amount, they tend to group together and become more prominent than when they are steady. In the sense that prominence can be considered equivalent to segregation, my arguments predict that FM can (and probably does) play a role in segregating a group of components from a background of other components. The only sense in which FM does not play a role in segregation is that its absence does not cause components to be segregated from each other in the way that, for example, the absence of harmonicity or of onset synchrony does. I think this distinction is an important one, and thank Dr McAdams for prompting me to make it more explicit.

[61]

Auditory segregation of competing voices: absence of effects of FM or AM coherence

QUENTIN SUMMERFIELD AND JOHN F. CULLING

MRC Institute of Hearing Research, University Park, Nottingham NG7 2RD, U.K.

SUMMARY

Four experiments sought evidence that listeners can use coherent changes in the frequency or amplitude of harmonics to segregate concurrent vowels. Segregation was not helped by giving the harmonics of competing vowels different patterns of frequency or amplitude modulation. However, modulating the frequencies of the components of one vowel was beneficial when the other vowel was not modulated, provided that both vowels were composed of components placed randomly in frequency. In addition, staggering the onsets of the two vowels, so that the amplitude of one vowel increased abruptly while the amplitude of the other was stationary, was also beneficial. Thus, the results demonstrate that listeners can group changing harmonics and can segregate them from stationary harmonics, but cannot use coherence of change to separate two sets of changing harmonics.

1. INTRODUCTION

There are at least two reasons for studying the auditory and perceptual processes which listeners use to attend selectively to one voice in a mixture of voices. First, speech is generally heard against a background of other sounds, including other voices. Thus an account of speech perception should include descriptions of the ways in which the elements of a voice are identified, grouped together, and separated from other sounds. Second, many hearing-impaired listeners have difficulty understanding speech in noise, particularly when the noise consists of other voices. Understanding the processes by which speech is normally extracted from interfering sounds, and the ways in which those processes break down in pathology, could lead to improved algorithms for speech enhancement in hearing aids.

The basic problem in segregating voices was set out by Broadbent & Ladefoged (1957): when two talkers speak concurrently, the spectrum of the sound reaching listeners' ears contains evidence of the formants of both voices. What cues enable listeners to assign each formant to the appropriate source? When both talkers produce voiced speech, the problem is that of correctly assigning each of the harmonics that define the formant peaks. Much work has sought to identify the mechanisms of spectral and temporal analysis that exploit the 'harmonicity' of the harmonics of a voice; i.e. the fact that the harmonics are found at frequencies that are integer multiples of their common fundamental frequency (F_0) (e.g. Broadbent & Ladefoged 1957; Darwin 1981; Scheffers 1983; Zwicker 1984; Gardner *et al.* 1989; Darwin & Culling 1990; Assmann & Summerfield 1990; Summerfield & Assmann 1991; Meddis & Hewitt 1992*b*). In this paper,

we are concerned with an additional issue: the role of time-varying cues. The experiments ask whether listeners can use correlated changes in the frequencies or amplitudes of harmonics, in addition to their harmonicity, to segregate competing voices.

2. EFFECTS OF COHERENT FREQUENCY MODULATION

When the fundamental frequency of a voiced vowel changes, the frequencies of its harmonics change coherently: they all rise or fall by the same percentage of their starting frequency. We shall refer to this example of common fate as 'coherent frequency modulation' (CFM) and ask: Can CFM help to group the harmonics of one voice and segregate them from the harmonics of a competing voice that are undergoing a different pattern of CFM? (For economy in writing, we shall describe voices that have different patterns of CFM as being 'incoherently modulated' and voices that have the same pattern of CFM as being 'coherently modulated'.) A demonstration by McAdams suggested that CFM might be a powerful grouping principle. He summed the waveforms of three synthetic vowels sung on different pitches. Applying CFM to the harmonics of one member of the triad caused it to stand out perceptually from the other two. Subsequent experiments (McAdams 1989; Marin & McAdams 1990) confirmed that CFM increased the perceptual prominence of one vowel in a mixture. However, its prominence was not affected by the status of the other two vowels. Prominence did not decrease when the other vowels were modulated coherently with the first, nor did it increase when the other vowels were modulated incoherently with the first. Thus McAdams concluded that CFM does not aid segregation. His

Phil. Trans. R. Soc. Lond. B (1992) **336**, 357–366
Printed in Great Britain

357

© 1992 The Royal Society and the authors

[63]

conclusion has been reinforced by the results of studies that have used accuracy of identification to measure the ability of listeners to segregate competing vowels or syllables (Chalikia & Bregman 1989; Gardner *et al.* 1989; Darwin & Culling 1990; Demany & Semal 1990). All show that a difference in F_0 is a potent cue for segregation, but that CFM does not make an independent contribution.

Carlyon (1991; see also this symposium) explained this outcome by arguing that listeners may not be able to use CFM. He demonstrated that listeners cannot distinguish coherent FM from incoherent FM carried on a small set of inharmonic tones. For example, in one experiment listeners were presented with tones at 400 Hz and 700 Hz, each modulated at a rate of 5 Hz with a modulation depth (zero-peak) of 5%. Listeners could not distinguish the case where the tones were modulated in phase from the case where their modulating waveforms were 180° out of phase. Carlyon argued that because listeners cannot detect whether components are modulated coherently or incoherently, they cannot be expected to use coherence of FM as a basis for grouping.

A different explanation is implicit in the writings of Chalikia & Bregman (1992). They argued that harmonicity is such a powerful cue for grouping that it permits all the segregation that can be achieved, leaving nothing for CFM to contribute. Chalikia & Bregman proposed that the way to demonstrate effects of CFM is to prevent harmonicity from playing a role, by synthesizing competing sounds whose components are placed randomly in frequency rather than harmonically. They carried out such an experiment. Inharmonic ('random') vowels were created from harmonic vowels by (i) randomly displacing harmonics in frequency within a circumscribed range and (ii) adjusting the amplitudes of the displaced harmonics to reinstate the original spectral envelope. In this way it was possible to convert a harmonic stimulus with a F_0 of, say, 100 Hz, into an inharmonic stimulus with a 'nominal F_0' of 100 Hz. A second inharmonic sound, whose nominal F_0 differed from the first by, say, 2 semitones, could then be created by changing the frequency of each component by 2 semitones.

Chalikia & Bregman used these procedures to generate pairs of inharmonic vowels whose nominal F_0s either rose or fell by 6 semitones over a duration of 2 s. In one set of pairs, the nominal F_0 values maintained a constant difference of 6 semitones. In another set, the initial and final differences were 6 semitones but the contours crossed. The members of the crossing pairs were identified slightly, but significantly, more accurately (91% correct compared to 87%) than the members of the parallel pairs, suggesting a small role for CFM.

Chalikia & Bregman's experiment was ingenious, but its implementation suffered from the problem that different stimuli, involving different ranges of nominal F_0, were presented in the different conditions. Thus, results could have been confounded by differences in the phonetic distinctiveness of the vowels depending on the precision with which components defined the locations of formant peaks. Experiment 1 sought to

distinguish Chalikia & Bregman's account of the role of CFM from Carlyon's account using a more rigorous psychophysical procedure.

3. MEASURING SEGREGATION THROUGH MASKING

We measure the effectiveness of cues for segregating voices using a masking procedure (Summerfield 1992) derived from procedures used by Demany & Semal (1990) and Summerfield & Assmann (1991). The procedure determines the minimal signal-to-noise ratio at which listeners can identify 'target' vowels in the presence of 'masking' vowels. An increased ability to segregate targets from maskers is revealed as a fall in the listener's masked threshold and can be quantified in dB. We use a two-interval, five-alternative forced-choice task. Maskers are presented in both intervals at a mean level of 60 dB (A). A target vowel is presented in one interval, chosen randomly. To score a correct response, listeners must indicate which interval contained the target and what its identity was. An adaptive staircase controls the target-to-masker ratio (TMR) and estimates the TMR giving 71% correct responses. The targets are exemplars of the five British–English vowels /a/, /i/, /ɜ/, /u/, and /ɔ/, with unchanging formant frequencies synthesized with a version of the cascade synthesizer described by Klatt (1980). The maskers are also five-formant sounds, but differ from the targets. To prevent listeners using unintended cues, two parameters are varied randomly between the intervals: overall level, so that an increase in loudness cannot be used to locate the interval containing the target; and the spectrum of the masker, so that the spectrum of the target cannot be recovered by computing the difference between the spectra of the sounds presented in the two intervals. Conditions are distinguished by changing the maskers not the targets. Thus, differences between the phonetic distinctiveness of the targets cannot confound the results. The maskers are drawn randomly from a set of ten. Thus, it is unlikely that listeners perform the task by learning the sound of each target combined with each masker.

Three experienced listeners with normal hearing took part in each experiment. Each listener provided two thresholds in each condition. Results are reported averaged over listeners. The test–retest reliability is such that differences between conditions of 3 dB are significant.

4. EXPERIMENT 1: EFFECTS OF CFM WITH HARMONIC AND INHARMONIC VOWELS

The experiment involved seven conditions which were distinguished by different relationships between the F_0 contours of maskers and targets. These relationships are shown schematically in the panels at the top of figure 1. Maskers and targets were 400 ms in duration with onsets and offsets shaped by 20 ms raised-cosine functions. The components of the targets were always modulated. Mean F_0s were chosen randomly from the set 100.0, 112.2, 126.0, and 141.4 Hz, whose members

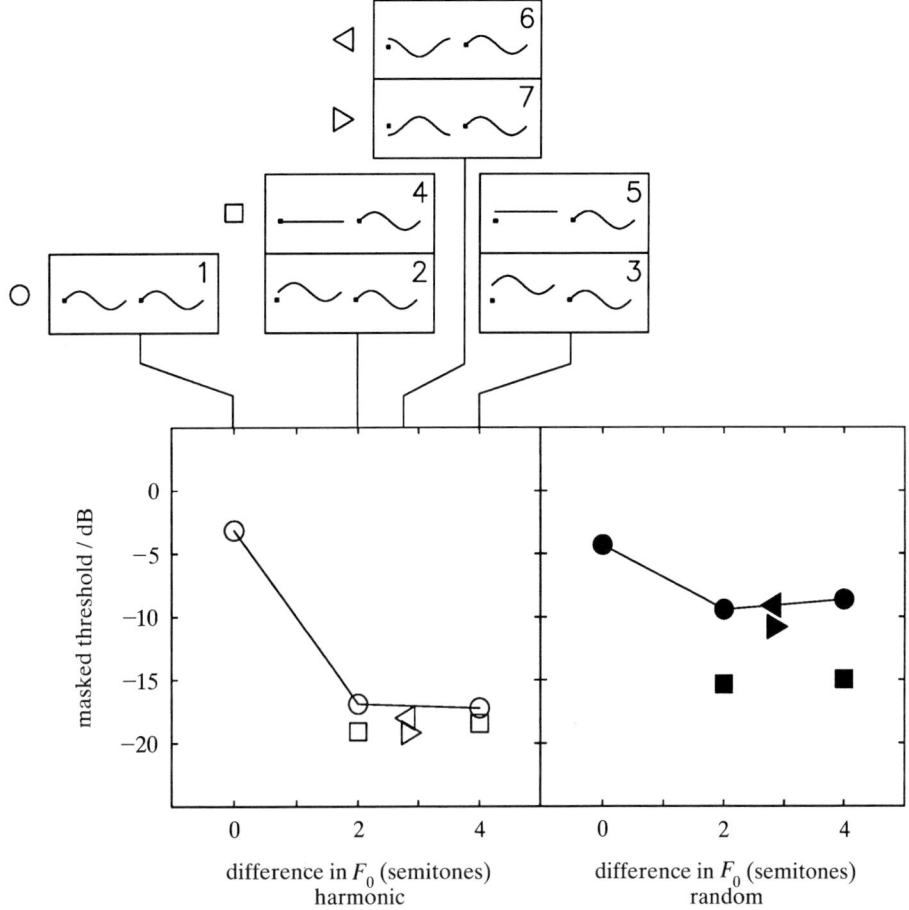

Figure 1. Results of experiment 1 for harmonic stimuli (open symbols, left graph) and inharmonic stimuli (filled symbols, right graph). Panels at the top of the plot illustrate the relationships between the F_0 contours of maskers (left trace in each panel) and targets (right trace). The small squares mark a constant point in time-frequency for reference. The numbers 1–7 identify the different conditions. Plotting symbols are shown beside each panel. Results in conditions with coherent modulation are shown by circles, in conditions with incoherent modulation by triangles, and in conditions where maskers were not modulated by squares.

are 2 semitones from their nearest neighbours. The modulation rate was 2.5 Hz and the modulation depth (zero-peak) was two semitones (12.2%). When maskers were modulated, their rate and depth were also 2.5 Hz and two semitones.

(a) *Role of* CFM *with harmonic stimuli*

Results obtained with harmonic stimuli are plotted in the left-hand panel of figure 1. Three effects can be seen. First, the circles show that thresholds fell by 14 dB when a difference of 2 semitones was introduced between maskers and targets (compare conditions [1] and [2]), but fell no further when the difference was increased to 4 semitones [3]. The result is compatible with earlier results showing that the benefits of F_0 differences reach a plateau at a difference of about 2 semitones (e.g. Summerfield & Assmann 1991). Second, thresholds were not significantly lower when targets were modulated against static maskers giving maximum differences in F_0 of 2 or 4 semitones (squares: conditions [4] and [5], respectively) than in the corresponding conditions where maskers and targets were both modulated with constant differences of 2 or 4 semitones (circles: conditions [2] and [3]).

Third, the triangles show that there was no significant advantage from modulating maskers and targets incoherently by either advancing [6] or retarding [7] the phase of the masker modulation by 90°. In these conditions, the maximum instantaneous difference in F_0 was 2.8 semitones. Thresholds were not significantly lower in conditions [6] and [7] (triangles) than in conditions [2] and [3] (circles) where maskers and targets were modulated coherently with constant differences of either 2 or 4 semitones. In other words, there was no advantage from giving maskers and targets different patterns of CFM over and above the advantage that would be expected from the maximum instantaneous difference in F_0 occurring during the modulation cycle. The results are compatible with the reports noted above which suggested that CFM plays no independent role in segregating harmonic sounds.

(b) *Effects of onset asynchronies with harmonic stimuli*

According to Chalikia and Bregman's explanation, no effect of CFM was shown in conditions [6] and [7] because the instantaneous differences in F_0 had already allowed thresholds to fall as far as they could in conditions [2] and [3]. We tested this aspect of their explanation by checking whether the introduction of an additional cue for segregation would cause thresholds to fall further. The further cue was a difference in onset time. Maskers started 200 ms before targets; they then continued for 400 ms and ended together. Staggering the onset times of competing sounds generally facilitates their segregation. For example, a harmonic that starts before the remaining harmonics in a complex makes a reduced contribution to the pitch of the complex (Darwin and Ciocca, 1991) and to the phonetic quality of a vowel (Darwin, 1984). Similarly, identification of the second vowel in a pair is more accurate if the second vowel starts after the first (Summerfield and Assmann, 1989). In the present experiments, we should expect thresholds to fall when maskers start before targets, unless differences in F_0 have already allowed thresholds to fall as far as they can.

The lengths of the open bars in Figure 2 show the amounts by which thresholds fell when maskers started 200 ms before targets, compared to the results plotted in Figure 1 where they started together. In the condition where there was no difference in F_0 between maskers and targets [1], thresholds fell significantly, showing that onset asynchrony can aid segregation. However, thresholds did not fall significantly when maskers and targets were modulated coherently with a constant difference of 2 semitones [2], nor in any of the other conditions. Thus, the outcome is compatible with the idea that no effect of CFM is found with harmonic stimuli because differences in F_0 allow thresholds to fall as far as they can. However, it is also possible that listeners simply cannot use CFM for segregation. The results obtained with the inharmonic stimuli, described in the next section, distinguish these alternatives.

(c) *Role of* CFM *with inharmonic stimuli*

The filled symbols in the right-hand panel of figure 1 show the results obtained with inharmonic stimuli. The circles show that thresholds fell by 5 dB when a difference of 2 semitones [2] or 4 semitones [3] was introduced between corresponding components in maskers and targets. The fall probably occurred for the following reasons. When maskers and targets had the same nominal F_0, they were composed of components with the same frequencies but different amplitudes and phases. Summation of the two waveforms distorted the spectral envelopes of both signals. Introducing a difference of 2 or 4 semitones between corresponding components reduced the interference, improving the definition of formant peaks in the targets. This outcome suggests that only 9 dB of the 14 dB fall obtained with harmonic stimuli in the analogous conditions should be attributed to effects of harmonicity.

The squares show that it is relatively easy to identify a modulated target against a static masker. In conditions [4] and [5] (squares), targets were modulated while maskers were static. Thresholds fell significantly compared to conditions [2] and [3] (circles) where

maskers and targets were both modulated. The result shows that a sound defined by changing components can 'stand out' against a background of static components.

However, the triangles show that a sound defined by one set of coherently modulated components does not stand out against a sound defined by components which are given a different pattern of CFM. Thresholds were not significantly lower when maskers and targets were modulated incoherently [6] and [7] (triangles) rather than coherently [2] and [3] (circles). Thus, CFM made no contribution to segregation beyond the contribution expected from the maximum instantaneous difference in (nominal) F_0 occurring during the modulation cycle.

(d) *Effects of onset asynchronies with inharmonic stimuli*

The solid bars in figure 2 show the effects of introducing a 200 ms onset asynchrony between inharmonic maskers and targets. In general, thresholds fell significantly when they were high in the synchronous case in figure 1 and failed to fall significantly when they started low, as in condition [4]. The crucial comparisons involve conditions [2], [6] and [7]. In conditions [2] and [6], thresholds fell significantly when the onset asynchrony was introduced. This result indicates that, if listeners had been sensitive to CFM, thresholds should also have fallen when FM incoherence was introduced. (This conclusion must be tempered slightly by the failure of condition [7] to show a significant fall.)

(e) *Summary*

Experiment 1 has demonstrated that CFM does not help listeners to segregate concurrent vowels. No advantage might be expected in the case of harmonic vowels because harmonicity allows all the segregation that can be achieved. However, CFM also failed to aid the segregation of inharmonic vowels. This outcome is compatible with Carlyon's (1991) conclusion that listeners cannot use CFM for segregation. It runs counter to Chalikia & Bregman's predictions.

Gardner *et al.* (1989) speculated that listeners have not included CFM in their armoury of grouping weapons because the uneven frequency responses of natural reverberant communication channels distort evidence of FM. More generally, Summerfield (1992) suggested that it is likely that CFM is not used because its exploitation would be computationally demanding, and is unnecessary on ecological grounds. To exploit CFM, listeners would have to track individual harmonics and compare their frequency contours. The problem might be soluble when only one source is present but could be intractable in the presence of a competing voice where each set of harmonics would have to be tracked across the changing background of the competing set. Instead, given the low incidence of natural sound sources generating discrete components at inharmonic frequencies, auditory analysis exploits harmonicity for grouping. Harmonicity can be exploited by

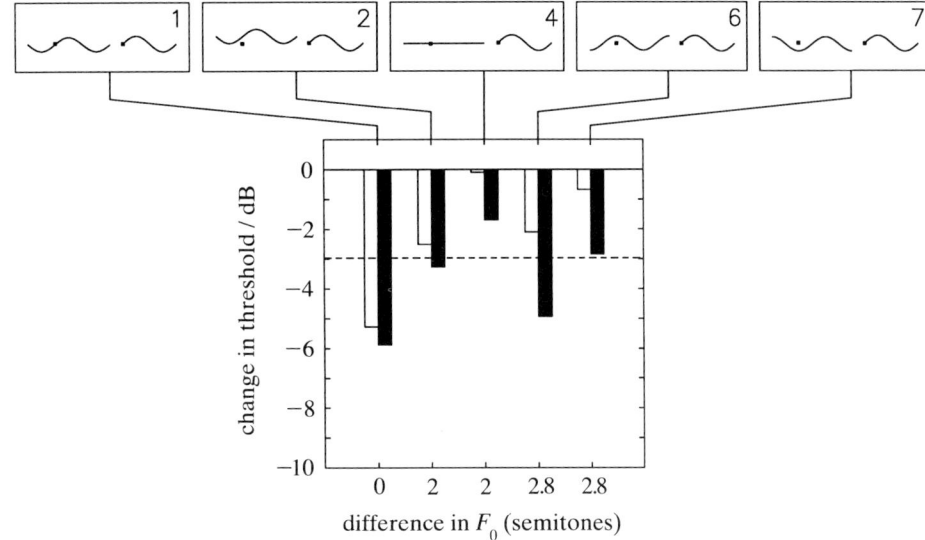

Figure 2. Effects of onset asynchrony in experiment 1 for harmonic stimuli (open bars) and random stimuli (filled bars). The dashed line shows the 3 dB difference between conditions required for significance. Panels at the top of the plot illustrate the relationships between the F_0 contours of maskers (left trace in each panel) and targets (right trace).

across-channel processes without the need to track individual harmonics (Assmann & Summerfield 1990; Meddis & Hewitt 1991)†. Moreover, as shown by the open symbols in figure 1 and the open bars in figure 2, sensitivity to CFM is unnecessary because harmonicity allows all the segregation that can be achieved.

The inability of listeners to use CFM for grouping could reflect a specific limitation in anlaysing the frequencies of individual harmonics, or a general difficulty in separating incoherently modulated sounds. Accordingly, the following experiments describe initial explorations of the ability of listeners to use coherent changes in amplitude to separate concurrent vowels.

5. EFFECTS OF COHERENT AMPLITUDE MODULATION

The experiments study effects of amplitude modulation in the sub-audio-frequency range which extends up to about 50 Hz. Such modulations are produced in speech by the control of air flow and acoustic radiation through the mouth by movements of the jaw, lips, and tongue. In the output of a bank of auditory filters, the modulations are found in a correlated form across a wide range of frequency channels. Their patterning can cue some phonetic distinctions (Rosen, this symposium) and their preservation in communication channels is important for intelligibility. Listeners use the modulations to group energy in different audio-frequency regions. For example, the release from masking demonstrated in co-modulation masking release (CMR) (Hall *et al.* 1984; Hall & Grose, this symposium) can be interpreted as a

---

† In some other circumstances listeners can track individual harmonics, because a harmonic that starts before the others in a complex makes a reduced contribution to the pitch of the complex (Darwin & Ciocca 1991) and its vowel colour (Darwin 1984).

consequence of grouping: the on-frequency band of noise and its flanking companion band are grouped together by virtue of their correlated patterning in amplitude, thereby allowing the unmodulated signal tone to be heard out from the on-frequency band. CMR has been viewed as a manifestation of mechanisms that segregate co-modulated speech formants from background noises.

A further experiment (Hall & Grose 1990) demonstrated that listeners can segregate concurrent signals carrying different patterns of AM. Hall & Grose measured CMR in conditions where the on-frequency band was centred on 1 kHz and six co-modulated flanking bands were centred on six multiples of 200 Hz around 1 kHz. Two 'co-deviant' bands centred on 900 Hz and 1100 Hz were introduced and modulated together but with a different envelope from the co-modulated bands. Their presence reduced the amount of release from masking compared with the release expected from the co-modulated bands. However, introducing six more co-deviant bands centred on other odd harmonics of 100 Hz around 1 kHz reinstated some of the lost CMR. Hall and Grose argued that increasing the number of co-deviant bands caused them to be grouped separately from the co-modulated bands and thereby prevented them from interfering with the unmasking effect. Although it is not clear whether it was important that the bands in each group were harmonically related, the result demonstrates that concurrent signals with different patterns of AM can be separated. A similar conclusion has been drawn from studies of 'modulation masking' (Bacon & Grantham 1989) which have demonstrated that AM at one rate (e.g. 8 Hz) can mask the detection of AM at the same rate more effectively than at other rates (e.g. 4 or 16 Hz; Houtgast 1989). It might be expected therefore that it should be easier to identify a target vowel in our paradigm if it is given a different pattern of amplitude modulation from the masking vowels.

However, other results make the outcome less certain. There is considerable evidence that it is difficult to make judgements about modulations in one frequency region if there are concurrent modulations occurring in different frequency regions. For example, compared with conditions where energy in different frequency regions is not modulated, when it is modulated it is harder to detect amplitude modulation itself (Yost & Sheft 1989) or to detect changes in the phase of modulation (Yost & Sheft 1989), the rate of modulation (Yost *et al.* 1989), or the depth of modulation (Moore *et al.* 1991; Moore & Shailer, this symposium). Moreover, the tuning of these effects of 'modulation detection interference' (MDI) to modulation rate is quite broad. As a result, Moore (1992) has suggested that across-channel masking may hinder the segregation of competing voices, despite the presence of short-term differences in modulation rate between the voices that might be expected to promote segregation.

From these results, it is difficult to predict whether it should be easier or more difficult to separate concurrent vowels if they are given different rates of AM rather than the same rate. Experiment 2 examined this issue.

6. EXPERIMENT 2: EFFECTS OF AMPLITUDE-MODULATION RATE?

Figure 3 shows the relationship between the intensity envelopes of maskers (thicker lines) and targets (thinner lines) in three of the five conditions. The modulation amplitude (zero-peak) of maskers and targets was 5 dB. The modulation rate of the targets was 8 Hz; that of the maskers ranged from 3.4 Hz to 19.0 Hz. A peak in the intensity envelope of the target coincided with a valley in the envelope of the masker half-way through each stimulus. Thus, if listeners can do no more than take advantage of the maximum instantaneous difference in level between maskers and targets, thresholds should be constant across the five conditions. Alternatively, if listeners can use a difference in

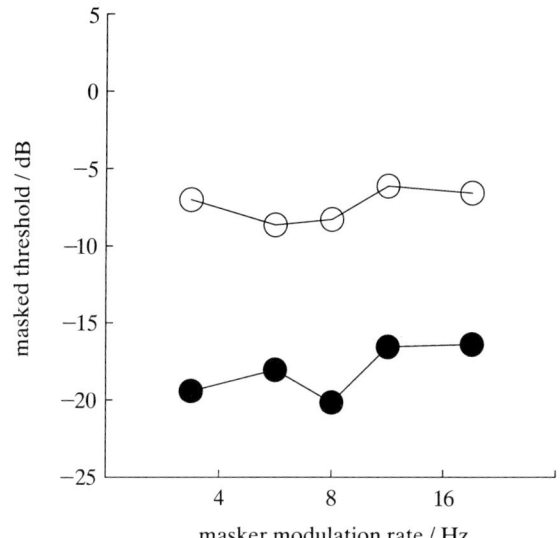

Figure 4. Results of experiment 2 for conditions in which the difference in F_0 between maskers and targets was 0 semitones (open circles) or 1 semitone (filled circles).

modulation rate to separate targets from maskers, thresholds should decline as the difference between the modulation rates increases. To obtain some generality, conditions were run in which maskers and targets had the same F_0 of 100 Hz and, separately, in which the F_0 of the maskers (105.9 Hz) was 1 semitone above the F_0 of the targets (100 Hz).

Mean thresholds from three listeners are shown in figure 4. As in experiment 1, there is a large effect of harmonicity; thresholds were 10–15 dB lower when maskers and targets possessed different F_0s. However, the results provide no evidence that listeners can use a difference in AM rate between two concurrently modulated 500 ms vowels to separate them perceptually. In fact, thresholds increased by 3–4 dB when maskers were given faster modulation rates than targets and had a different F_0. Thus, introducing a difference in AM rate slightly disrupted segregation.

7. EXPERIMENTS 3 AND 4: EFFECTS OF AMPLITUDE-MODULATION PHASE?

The modulating waveforms of maskers and targets in experiment 2 had different phases, even when they had the same 8 Hz rate. Thus, targets and maskers were always modulated incoherently. The incoherence itself might have permitted a material amount of segregation since the formants of one vowel were rising in amplitude at times when the formants of the other were falling. Accordingly, Experiment 3 asked whether a difference in modulator phase aids segregation when maskers and targets are modulated at the same rate.

Maskers and targets had a duration of 400 ms and were modulated at a rate of 2.5 Hz either coherently (in-phase) or incoherently (by advancing the phase of the masker modulation by 180°). Modulation depth (zero-peak) was varied from 1 dB to 5 dB. The first line of figure 5, labelled 'condition 1', shows the relationship between the amplitude envelopes of

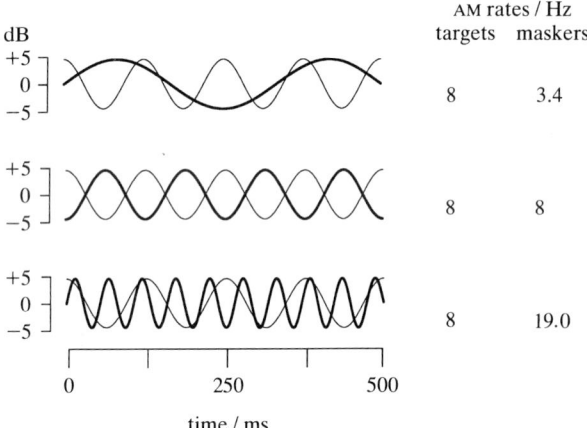

Figure 3. Amplitude envelopes of a subset of the maskers (thicker lines) and targets (thinner lines) used in experiment 2.

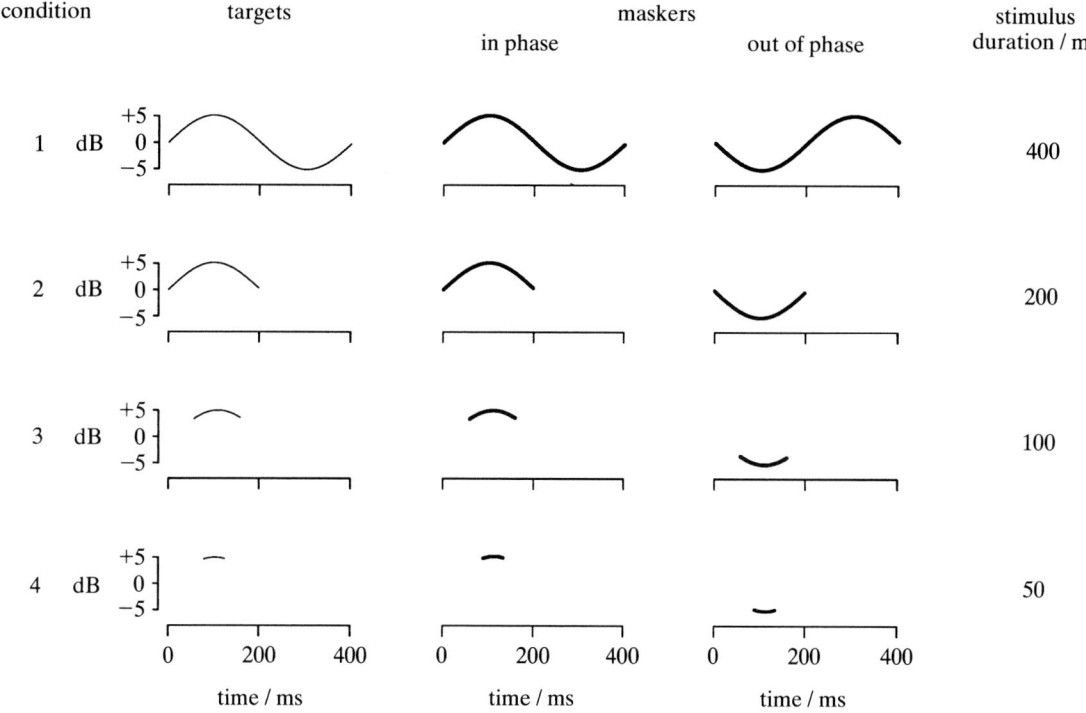

Figure 5. Amplitude envelopes of maskers (thicker lines) and targets (thinner lines). Experiment 3 involved condition 1 only. Experiment 4 involved all four conditions.

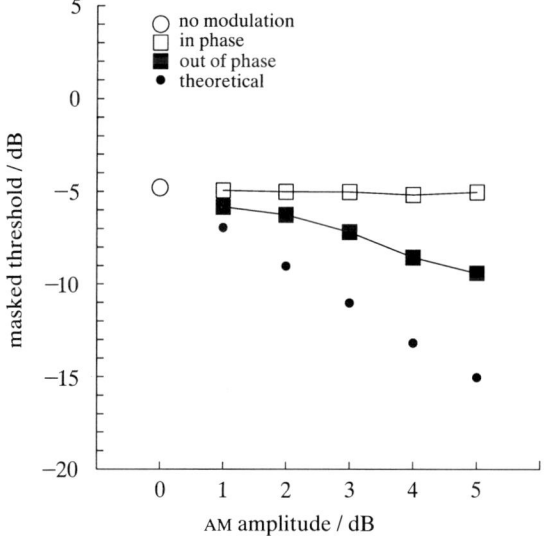

Figure 6. Results of experiment 3.

maskers and targets in the in-phase and out-of-phase conditions for the case where the modulation depth was 5 dB.

The experiment was intended to establish whether AM coherence plays an independent role in segregating voices. In other words, do the benefits of incoherent modulation exceed the advantages expected from the maximum instantaneous difference in level between maskers and targets that occurs during the combined stimulus?

Figure 6 shows thresholds obtained with in-phase stimuli (open squares) and out-of-phase stimuli (filled squares) as modulation depth increased from 1 dB to

5 dB. Thresholds were constant in the in-phase condition, but fell in the out-of-phase condition. The small filled circles have been plotted below the thresholds obtained in the in-phase condition (open squares) by an amount equal to the peak-to-peak modulation depth. Thus, the small circles plot the thresholds that would have occurred in the out-of-phase condition if listeners could take advantage of the maximum instantaneous difference in level between maskers and targets that occurred during the combined stimulus. In fact, the thresholds measured in the out-of-phase condition (filled squares) are higher than these theoretical points. Thus, not only could listeners not take advantage of AM incoherence, they could not even take advantage of the maximum instantaneous difference in level between maskers and targets.

A possible explanation for this result is based on the idea that a shorter duration of the targets was detectable in the out-of-phase condition than in the in-phase condition. Consider condition 1 in figure 5 again. In the in-phase condition, the local TMR is constant throughout the 400 ms duration of the combined stimulus. In the out-of-phase condition, in comparison, the local TMR varies over the duration of the stimulus. It is maximal, momentarily, at the point 100 ms after the start of the stimulus and is minimal at the 300 ms point. Imagine that a target is added to a masker at the same overall TMR in both the in-phase and out-of-phase conditions. The local TMR in the out-of-phase condition would be higher at the 100 ms point than it would be at any point in the in-phase condition. Hence, as was observed in figure 6, thresholds would be expected to be lower in the out-of-phase condition (filled squares) compared to the in-phase condition (open squares). Now consider the situation that would arise if the overall TMR was reduced until the local TMR at the 100 ms point in the out-of-phase condition equalled the constant TMR observed at threshold in the in-phase condition. In this situation listeners would be able to detect only a brief segment of the target in the out-of-phase condition close to the 100 ms point in the stimuli. This segment would obviously be shorter than the 400 ms segment that could be detected in the in-phase condition. It is well established that performance in tasks requiring detection or discrimination deteriorates as the duration of the stimulus is reduced (e.g. Viemeister & Wakefield 1991). Hence, thresholds observed in the out-of-phase condition (filled squares in figure 6) would be higher than those predicted (small filled circles) from performance in the in-phase condition. In essence, listeners would not achieve predicted performance because they would be basing their judgements on a shorter effective duration of the targets in the out-of-phase condition than in the in-phase condition. We shall refer to this explanation as the 'time-intensity trading' account of the results of experiment 3.

Experiment 4 sought to verify the 'time-intensity trading' account and to establish whether another factor, described below, might also have played a role. The experiment compared performance in all four of the conditions illustrated in figure 5. Moving from condition 1 to condition 4, the stimuli were progressively restricted to a 50 ms segment centred on the point where the local TMR is maximal in the out-of-phase condition. The rationale is as follows. Suppose that the time-intensity trading account holds. In which case, in the out-of-phase conditions of experiment 3 listeners would have based their judgements on a brief segment of the target close to the 100 ms point in the stimuli. Let that segment have a duration of D ms. Thus, in experiment 4, thresholds should remain constant in the out-of-phase condition until the duration of the stimuli is reduced to a value less than D. In the in-phase conditions of experiment 3, in comparison, listeners could accumulate evidence of the targets over the full duration of the stimulus. Thus, here in the in-phase conditions of experiment 4, performance should suffer as stimulus duration is reduced and thresholds should rise.

The results‡ of experiment 4, shown in figure 7, are compatible with these predictions. As the duration of the stimuli was reduced, thresholds rose in the in-phase conditions (open squares) but stayed constant in the out-of-phase conditions (filled squares). The constancy of the thresholds in the out-of-phase conditions suggests that the duration D could be as short as 50 ms.

An additional factor which might have affected the results of experiment 3 is that listeners may not have known 'when to listen'. Because of MDI, or for some other reason, they might not have been able to select the moment in the stimuli when the TMR was

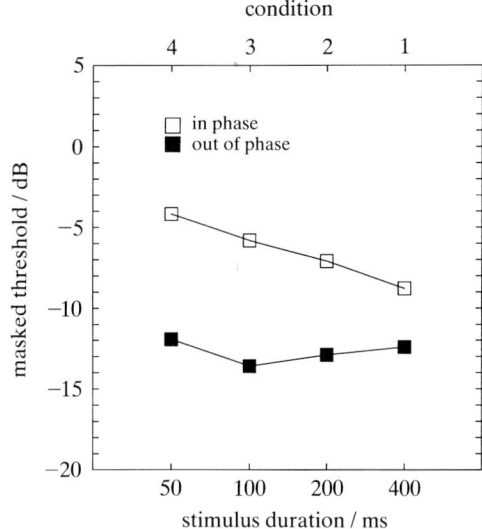

Figure 7. Results of experiment 4.

maximal. Instead, they might have averaged evidence of the target over the part of the stimulus giving a reasonably good TMR. If so, performance in the out-of-phase conditions of experiment 4 might have been expected to improve as the duration of the stimuli was reduced, because listeners would be able to focus on the moment giving the maximal TMR. However, performance did not improve as stimulus duration was reduced (filled squares in figure 7). Thus, there is no evidence that MDI, or any other process mediated specifically by AM, significantly limited performance in experiment 3. Rather, the results of that experiment are explained by the time-intensity trading account.

In summary, experiments 3 and 4 have shown that segregation of concurrent vowels is not facilitated by incoherent amplitude modulation. Establishing the generality of these conclusions, however, requires further experiments using faster modulation rates than the relatively slow (2.5 Hz) rate used here.

8. CONCLUSIONS

We have found no evidence that listeners can use coherent changes in the frequencies or amplitudes of the harmonics of a vowel to separate that vowel from a competing vowel whose harmonics are undergoing a different pattern of modulation. Neither form of 'common fate' was useful when both vowels were modulated. However, there were benefits when the components of one vowel were modulated in frequency while the components of the other were stationary. In addition, staggering the onsets of the vowels, so that one underwent an abrupt increase in amplitude while the other was static, was also beneficial. Thus, there is evidence that certain types of change in frequency or amplitude can help segregate the changing vowel from a static one. However, even these benefits were small in relation to the benefits from an absolute difference in F_0 between the vowels. The potency of harmonicity in relation to other cues for voice segregation has also been noted by Shackle-

---

‡ The 4 dB difference in overall performance level between experiments 3 and 4 can be attributed to the participation of different subjects in the two experiments.

ton *et al.* (1991) who compared its benefits with those of binaural cues. Together, these results reinforce the idea (e.g. Stubbs & Summerfield 1991) that signal-processing approaches to voice segregation should exploit harmonicity as one of the primary cues.

We thank Mark Haggard, Chris Darwin and Bob Carlyon for constructive comments on drafts of this paper.

REFERENCES

Assmann, P.F. & Summerfield, Q. 1990 Modeling the perception of concurrent vowels: vowels with different fundamental frequencies. *J. acoust. Soc. Am.* **88**, 680–697.

Bacon, S.P. & Grantham, D.W. 1989 Modulation masking: effects of modulation frequency, depth, and phase. *J. acoust. Soc. Am.* **85**, 2575–2580.

Broadbent, D.E. & Ladefoged, P. 1957 On the fusion of sounds reaching different sense organs. *J. acoust. Soc. Am.* **29**, 708–710.

Carlyon, R.P. 1991 Discriminating between coherent and incoherent frequency modulation of complex tones. *J. acoust. Soc. Am.* **89**, 329–340.

Chalikia, M.H. & Bregman, A.S. 1989 The perceptual segregation of simultaneous auditory signals: Pulse train segregation and vowel segregation. *Percept. Psychophys.* **46**, 487–496.

Chalikia, M.H. & Bregman, A.S. 1992 The perceptual segregation of simultaneous vowels with harmonic, shifted, and random components. (In preparation.)

Darwin, C.J. 1981 Perceptual grouping of speech components differing in fundamental frequency and onset time. *Q. Jl exp. Psychol.* **33**A, 185–207.

Darwin, C.J. 1984 Perceiving vowels in the presence of another sound: constraints on formant perception. *J. acoust. Soc. Am.* **76**, 1636–1647.

Darwin, C.J. & Ciocca, V. 1992 Grouping in pitch perception: effects of onset asynchrony and ear of presentation. *J. acoust. Soc. Am.* (In the press.)

Darwin, C.J. & Culling, J.F. 1990 Speech perception seen through the ear. *Speech Communic.* **9**, 469–475.

Demany, L. & Semal, C. 1990 The effect of vibrato on the recognition of masked vowels. *Percept. Psychophys.* **48**, 436–444.

Gardner, R.B., Gaskill, S.A. & Darwin, C.J. 1989 Perceptual grouping of formants with static and dynamic differences in fundamental frequency. *J. acoust. Soc. Am.* **85**, 1329–1337.

Hall, J.W., Haggard, M.P. & Fernandes, M. 1984 Detection in noise by spectro-temporal pattern analysis. *J. acoust. Soc. Am.* **76**, 50–57.

Hall, J.W. & Grose, J.H. 1990 Comodulation masking release and auditory grouping. *J. acoust. Soc. Am.* **88**, 119–125.

Houtgast, T. 1989 Frequency selectivity in amplitude-modulation detection. *J. acoust. Soc. Am.* **85**, 1676–1680.

Klatt, D.H. 1980 Software for a cascade/parallel formant synthesizer. *J. acoust. Soc. Am.* **67**, 971–995.

Marin, C.M.H. & McAdams, S. 1991 Segregation of concurrent sounds. II: Effects of spectral envelope tracing, frequency modulation coherence, and frequency modulation width. *J. acoust. Soc. Am.* **89**, 341–351.

McAdams, S. 1989 Segregation of concurrent sounds. I: Effects of frequency modulation coherence. *J. acoust. Soc. Am.* **86**, 2148–2159.

Meddis, R. & Hewitt, M.J. 1991 Virtual pitch and phase sensitivity studied using a computer model of the auditory periphery: Pitch identification. *J. acoust. Soc. Am.* **89**, 2866–2882.

Meddis, R. & Hewitt, M.J. 1992 Modelling the identification of concurrent vowels with different fundamental frequencies. *J. acoust. Soc. Am.* **90**, 233–245.

Moore, B.C.J. 1992 Across-channel masking and co-modulation masking release. In *Audition, speech, and language* (ed. M. E. H. Schouten). Berlin: Mouton. (In the press.)

Moore, B.C.J., Glasberg, B.R., Gaunt, T. & Child, T. 1991 Across-channel masking of changes in modulation depth for amplitude- and frequency-modulated signals. *Q. Jl exp. Psychol.* **43**A, 327–348.

Scheffers, M.T.M. 1983 Sifting vowels: auditory pitch analysis and sound segregation. Ph.D. thesis, University of Groningen, The Netherlands.

Shackleton, T.M., Meddis, R. & Hewitt, M.J. 1992 The role of binaural cues in the identification of simultaneously presented vowels. *Q. Jl exp. Psychol.* (Submitted.)

Summerfield, Q. 1992 Roles of harmonicity and coherent frequency modulation in auditory grouping. In *Audition, speech, and language* (ed. M. E. H. Schouten). Berlin: Mouton. (In the press.)

Summerfield, Q. & Assmann, P.F. 1989 Auditory enhancement and the perception of concurrent vowels. *Percept. Psychophys.* **45**, 529–536.

Summerfield, Q. & Assmann, P.F. 1991 Perception of concurrent vowels: effects of harmonic misalignment and pitch-period asynchrony. *J. acoust. Soc. Am.* **89**, 1364–1377.

Summerfield, Q., Sidwell, A. & Nelson, T. 1987 Auditory enhancement of changes in spectral amplitude. *J. acoust. Soc. Am.* **81**, 700–708.

Stubbs, R.J. & Summerfield, Q. 1991 Effects of signal-to-noise ratio, signal periodicity, and degree of hearing impairment on the performance of voice-separation algorithms. *J. acoust. Soc. am.* **89**, 1383–1393.

Viemeister, N.F. & Wakefield, G.H. 1991 Temporal integration and multiple looks. *J. acoust. Soc. Am.* **90**, 858–865.

Yost, W.A. & Sheft, S. 1989 Across-critical-band processing of amplitude-modulated tones. *J. acoust. Soc. Am.* **85**, 848–857.

Yost, W.A., Sheft, S. & Opie, J. 1989 Modulation interference in detection and discrimination of amplitude modulation. *J. acoust. Soc. Am.* **86**, 2138–2147.

Zwicker, U.T. 1984 Auditory recognition of diotic and dichotic vowel pairs. *Speech Communic.* **3**, 265–277.

Discussion

A. J. Fourcin (*Department of Phonetics and Linguistics, University College London, U.K.*). One of the findings reported was that the listener makes no use of fundamental frequency contour – intonation related information – in segregating the outputs of two competing 'speakers'. This result may perhaps be modified if speech-like patterns are used. Although the normal listener has an excellent intrinsic knowledge of the intonation patterning of normal spoken language and can use this as a basis for speaker identification, phrase-level sinusoidal fundamental frequency or intonation contours will be foreign to his or her experience. A useful extension to the present experiments might come from the use of contours which are based on real utterances. These can be manipulated, so that they are of experimentally convenient centre frequency and range, but conserved in respect of their idiosyncratic shapes (a technique which has been shown to preserve speaker identity information (Abberton & Fourcin 1978)).

A similar possibility for exploring the utility of using more natural, listener-experienced, stimuli comes from the employment of amplitude contour information which is speech derived. This also has been shown to be a source of speaker identification information (Atal 1968).

In both of these cases, the prediction is that the enhanced use of cognitive constraints coming from prior speech knowledge will enable listeners better to disentangle competing speech stimuli. More generally, it may always prove advantageous in exploring speech perceptual processing to pay close attention to the structure of speech itself.

References

Abberton, E. & Fourcin, A. 1978 Intonation and speaker identification. *Lang. Speech* **21**, 305–318.

Atal, B.S. 1968 Automatic speaker recognition based on pitch contours. Ph.D. thesis Polytechnic Institution of Brooklyn.

Q. SUMMERFIELD. It is important to distinguish low-level processes, which are used to group the harmonics of a single voice, from higher-level processes, which ensure that the groups of harmonics created by the low-level analyses are linked appropriately over time. A related distinction has been drawn by Bregman (1991) between primitive and schema-based grouping principles. Our experiments concern primitive processes reflecting physical constraints. Professor Fourcin's comment concerns schema-driven processes reflecting linguistic constraints.

The distinction can be illustrated by considering the computational problem faced by a system which attempts to separate sentences spoken concurrently by two talkers. A first step could be to locate harmonics. The next step would be to group the harmonics into two sets, one for each talker. The cues that might be used to do this include the following primitive grouping principles: (i) harmonicity: components whose frequencies are multiples of a common fundamental should be grouped together; (ii) coherent frequency modulation: components whose frequencies change in

the same direction by the same percentage of their starting frequency should be grouped together; (iii) onset–offset synchrony: components that start and stop at the same time should be grouped together; (iv) coherent amplitude modulation: components whose amplitudes rise or fall coherently should be grouped together; and (v) concurrent change: components whose frequencies are changing should be grouped separately from components whose frequencies are static. The experiments described in our paper show that factors (i), (iii), and (v) can be used by listeners. We did not find evidence that listeners could use factors (ii) or (iv).

At this stage, the system has formed two groups of components at each of a succession of moments in time. The next problem is to string together the appropriate members of each group. For example, suppose that at time t_1 the system has established that two groups of harmonics are present, with F_0s of (i) 150 and (ii) 190 Hz, while at time t_2, there are two groups with F_0s of (iii) 170 and (iv) 160 Hz. The task now is to establish whether group (iii) is the continuation of group (i) or group (ii). It is at this stage that the linguistic schema-based principles of the type mentioned by Professor Fourcin are likely to play a role. For example, continuity of pitch would help to solve the problem of grouping (iii) with (i) or (ii), particularly if the continuity accorded with linguistic rules.

We believe that the best way to study the primitive principles is to use heavily constrained stimuli. Clearly, however, more natural stimuli should be used to study the schema-based linguistic principles, as Professor Fourcin suggests. By using the two approaches, Brokx & Nooteboom (1982) demonstrated that both primitive and schema-based principles play a role when listeners are required to identify words in sentences spoken by competing talkers.

References

Bregman, A.S. 1991 *Auditory scene analysis: the perceptual organisation of sound.* Cambridge, Massachusetts: MIT Press.

Brokx, J.P.L. & Nooteboom, S.G. 1982 Intonation and the perceptual separation of simultaneous voices. *J. Phonet.* **10**, 23–36.

Temporal information in speech: acoustic, auditory and linguistic aspects

STUART ROSEN

Department of Phonetics and Linguistics, University College London, 4 Stephenson Way, London NW1 2HE, U.K.

SUMMARY

The temporal properties of speech appear to play a more important role in linguistic contrasts than has hitherto been appreciated. Therefore, a new framework for describing the acoustic structure of speech based purely on temporal aspects has been developed. From this point of view, speech can be said to be comprised of three main temporal features, based on dominant fluctuation rates: envelope, periodicity, and fine-structure. Each feature has distinct acoustic manifestations, auditory and perceptual correlates, and roles in linguistic contrasts. The applicability of this three-featured temporal system is discussed in relation to hearing-impaired and normal listeners.

1. INTRODUCTION

One of the most important properties of the normal auditory mechanism is that it acts as a frequency analyser. Therefore, when exploring the relationship between the perceptual attributes of speech sounds and their acoustic structure, most emphasis is placed on the frequency spectrum. So, for example, we talk about the frequencies of the formants in a vowel, or the multi-harmonic nature of voiced speech. Recently, however, there has been much greater interest in the purely temporal properties of speech sounds, for three main reasons.

First, from psychoacoustical studies, there is now a general consensus that place–frequency mechanisms on their own can not account for many aspects of the perception of pitch, and by implication the perception of intonation in speech (for relevant reviews, see Moore & Glasberg (1986); Rosen & Fourcin (1986)). Further confirmation of the importance of temporal features in pitch perception comes from Assmann & Summerfield's (1990) attempts to model the perception of concurrent vowels which have different fundamental frequencies. They found that only models which used time, as well as place, information, could account even reasonably well for the performance of human listeners. There is also good evidence from studies on the perception of phase manipulations that temporal factors play a role even in the perception of timbre, an attribute most commonly associated with the spectral shape of a sound (see Darwin & Gardner 1986; Patterson 1987; Rosen 1986, 1987).

Second, theoretical models derived from physiological evidence suggest that temporal information is important both for the perception of melodic pitch and for the auditory representation of spectral shape (Sachs & Miller 1985; Sachs et al. 1983). Finally, and perhaps most strikingly, there is the evident success of the large number of patients who have received single-channel cochlear implants. Such systems deliver an electrical signal based on the speech waveform to a single electrode placed in or near the cochlea, thus allowing no place-based frequency analysis. Yet, many of these patients have performed surprisingly well, even to the extent of being able to understand unknown sentences on the basis of an auditory signal alone (Hochmair-Desoyer et al. 1980, 1985). Furthermore, it appears that a number of implanted children are able to perform significantly better than any user implanted as an adult, presumably due to the greater neuronal plasticity of children (Luxford et al. 1987). These results have stimulated hypotheses about the extent to which purely temporal information can be effective in speech perception, and led to experiments in which normally hearing listeners are asked to identify speech sounds processed so as to contain temporal variations, but not spectral ones (Van Tasell et al. 1987; Rosen 1989). There has, however, been little detailed consideration of the temporal structure of speech (as displayed directly in the waveforms of figure 1) and its relation to various linguistic contrasts.

2. A FRAMEWORK FOR DESCRIBING TEMPORAL INFORMATION IN SPEECH

Therefore, as a complement to the standard Fourier-based spectral approach, we have developed a three-way partition of the temporal structure of speech based on dominant temporal fluctuation rates. Without implying anything about the extent to which this system is generally applicable (in particular to normal listeners), we now describe the three features – envelope, periodicity and fine-structure – in relation to their acoustic, auditory and perceptual and linguistic correlates. In a later section, we discuss the appro-

Phil. Trans. R. Soc. Lond. B (1992) **336**, 367–373
Printed in Great Britain

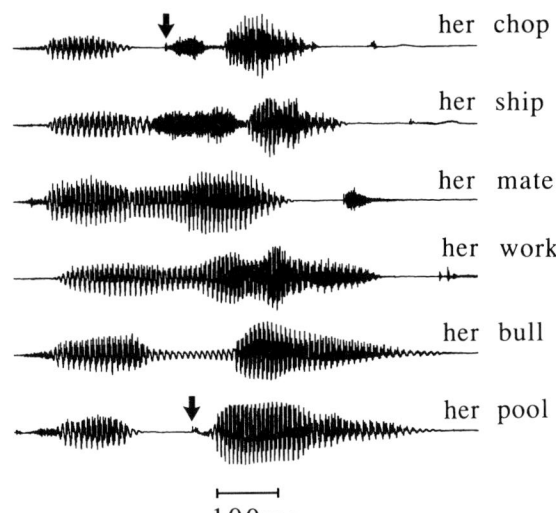

100 ms

Figure 1. Speech pressure waveforms of six phrases uttered by a male adult speaker. As postvocalic 'r' is not pronounced in the speech of this speaker (as is generally the case for so-called Received Pronunciation), the initial consonants of the second word in each phrase are in an inter-vocalic position. For later reference, the two arrows (in 'chop' and 'pool') indicate the release bursts typical of plosive type sounds.

priateness of using a temporal feature system in describing the speech-perceptual performance of various categories of hearing-impaired and normal listener.

(a) *Envelope*

We will refer to fluctuations in overall amplitude at rates between about 2 and 50 Hz as envelope information†. Variously known as 'amplitude envelope', 'time–amplitude', or 'time–intensity' information, this is typically what is meant by 'temporal information' in much of the literature. Envelope may be described mainly by such acoustic features as intensity, duration, rise time and fall time. Its main auditory correlates are loudness, length, attack and decay. Envelope's low frequency variations can convey four main types of linguistic information.

(i) *Segmental cues to manner of articulation*

Consider, for example, the voiceless affricate–fricative distinction (/tʃ/ and /ʃ/, contrasting the initial consonants of 'chop' and 'ship') for which a number of envelope features are known to be influential (Dorman *et al.* 1980; Gerstman 1957; Howell & Rosen 1983, 1987; Repp *et al.* 1978). The frication noise of /tʃ/ has a quicker rise time and shorter overall duration than the corresponding frication of /ʃ/. /tʃ/ has a short release burst whereas /ʃ/ does not. Short release bursts (transients) typically indicate plosive type sounds. Silent gaps too may indicate the presence of a voiceless plosive or affricate (as in the intervocalic

† Note that this 'envelope' is distinct from, although related to, the 'envelope' derived through use of the analytic signal. In the terms used here, the 'envelope' derived from the analytic signal typically contains both envelope and periodicity information (see Seggie 1986*a*, *b*).

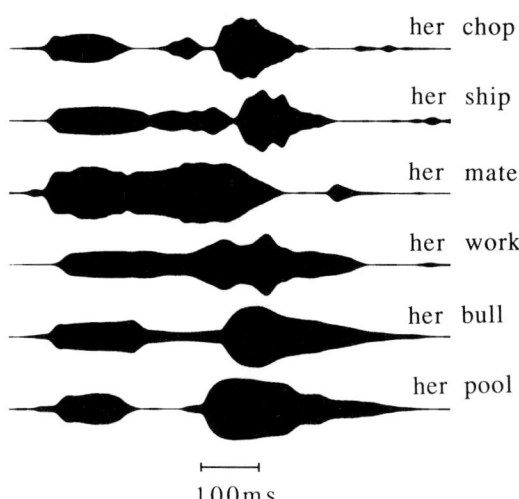

100 ms

Figure 2. Waveforms obtained from those of figure 1 by full-wave rectification and low-pass filtering at 20 Hz. This preserves much of the envelope information, but eliminates that occurring at relatively high fluctuation rates. Note, for example, the loss of the release bursts of /tʃ/ in 'chop' and /p/ in 'pool' evident in figure 1.

/tʃ/ and /p/ of figure 2: Bailey & Summerfield 1980; Summerfield *et al.* 1981). More generally, it has been proposed that relatively fast changes in overall amplitude mark consonants from non-consonants (Stevens 1980, 1981; Stevens & Blumstein 1981), or continuants from non-continuants (Shinn & Blumstein 1984).

(ii) *Segmental cues to voicing*

Voiced sonorants (vowels, semivowels, nasals and laterals) typically have greater amplitudes than voiceless obstruents (as in the /m/ of 'mate' versus the /p/ of 'pool' in figure 2). The existence and duration of silent intervals may be important in distinguishing voiced from voiceless plosives in intervocalic position (as in the figured /b/ of 'bull' versus /p/ of 'pool'). In some contexts, vowel duration, in so far as it is cued by envelope, can give information about voicing in the following consonant (Umeda 1975). Generally speaking, however, the envelope cues to voicing contrasts are weak.

(iii) *Segmental cues to vowel identity*

The duration of vowels varies lawfully with vowel quality, and so can signal some information about it, albeit weakly. Many languages use duration contrastively (along with changes in quality) to distinguish among vowels (see Lehiste (1970, pp. 18–19, 30–35) for a review). For example, other things being equal, the vowel in 'heed' tends to be of significantly longer duration than that of 'hid'.

(iv) *Prosodic cues*

Dynamic envelope cues can be used to assist syllabification (as Mermelstein (1975) has shown in an automated procedure), and relative amplitude (on a more static basis) probably plays a minor role in the assignment of stress in words (for example in distinguishing the verb 'rebel' from the noun 'rebel': see Crystal (1969, pp. 113–120); Fry (1968): Lehiste (1970, pp. 36–38, 120–139) for reviews of the relevant literature). In so far as amplitude onsets and offsets can demarcate linguistic units (vowel, syllable or word), much information about duration, and hence speech rhythm and tempo can also be extracted from envelope cues. Duration itself appears to play a role in word-level stress (for example, as in 'rebel' versus 'rebel') whereas information about tempo could assist listeners in normalizing for speech rate variations in segmental (Miller 1981) and prosodic contrasts. Variations in speech rate can also carry distinctions in meaning (Crystal 1969, pp. 152–156) or indicate parenthetical comments.

(b) Periodicity

Properties of the speech signal which relate to the distinction between periodic and aperiodic stimulation, and to the rate of periodic stimulation, shall be referred to simply as periodicity information. Periodic sounds fluctuate primarily at rates between about 50 and 500 Hz, whereas aperiodic sounds typically fluctuate at rates from a few kHz up to 5–10 kHz (although they can fluctuate at rates below 1 kHz). Because periodic and aperiodic sounds generally differ so greatly in their characteristic rates of fluctuation, it may be useful at times to think of periodicity information as being divided into two subclasses (periodic and aperiodic) with different dominant fluctuation rates, but which both give information about the source of excitation in speech production. Note, however, that the ability to distinguish periodic from aperiodic stimulation does not necessarily depend upon differential sensitivity to high fluctuation rates (that is, the ability to distinguish among sounds which fluctuate at different high rates). It is enough to distinguish absence of stimulation (indicating silence) from stimulation with low-frequency periodicity (below, say, 1 kHz, indicating a periodic sound) from stimulation of a high (but otherwise indeterminate) fluctuation rate (indicating an aperiodic sound). Similarly, detection of aperiodicity may rely on the irregular nature of the low-frequency fluctuations in a rapidly fluctuating sound.

The acoustic contrast of periodicity versus aperiodicity is reflected in the time domain as regularity versus irregularity of the speech signal, and in the frequency domain as the distinction between a harmonic and a continuous spectrum. Auditorily, this contrast is perceived as one of buzziness versus noisiness. Changes in the rate of periodic fluctuations are reflected in changes of fundamental frequency, which lead to a sensation of melodic pitch. Periodicity information directly conveys two main types of linguistic information.

(i) Segmental information about voicing and manner

The presence of low-frequency quasi-periodic acoustic energy in a speech signal is a reflection of the quasi-periodic vibrations of the vocal folds (as in the /m/ of mate in figure 3). Such sounds are said to be

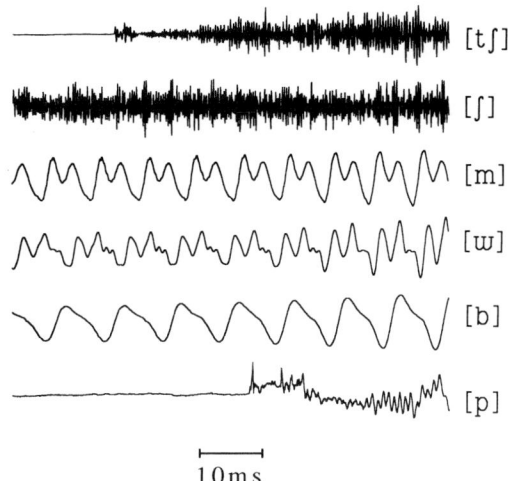

Figure 3. Sections of the speech pressure waveforms of figure 1, chosen so as to illustrate periodicity information. Note the irregular waveforms associated with the voiceless sounds /tʃ/, /ʃ/ and /p/, and the regularity exhibited by the quasi-periodic voiced sounds /m/, /w/ and /b/.

voiced, and such voicing is the most important cue to the phonological feature of voicing, perhaps the most basic distinction in all of the world's languages. Similarly, in many languages (such as English and French) there is an association between manner and voicing features (e.g. all nasals are voiced) which permits manner information to be obtained from information about phonetic voicing patterns. Speech segments which are aperiodic result from turbulence noise generated by aerodynamic flow between closely spaced articulators. Such aperiodicity can be a strong cue for voicelessness, or to the fricative manner of articulation (as in the /ʃ/ of 'ship' in figure 3).

(ii) Prosodic information relating to intonation and stress

The fundamental frequency of quasi-periodic energy in a speech signal is a reflection of the rate of vocal fold vibration, and is the prime acoustic correlate of the perception of voice pitch. Linguistically meaningful patterns of voice pitch are known as intonation and tone, and play important roles in accenting syllables in words and sentences, in clarifying ambiguous pronoun references, in marking syntactic units and in distinguishing questions and statements (for reviews see Fry (1968); Lehiste (1970); Rosen & Fourcin (1986)). Furthermore, in tone languages like Chinese, voice pitch patterns have a lexical function: that is, they distinguish different dictionary meanings of a word. For example, in Cantonese Chinese, the syllable 'yee' may mean 'clothes', 'chair', 'meaning', 'child', 'ear' or 'two', depending upon the pitch contour used when uttering it. Even English has a minor instance of this, in that voice pitch contours can play an important role in distinguishing between the verbal and nominal function of a word (as in 'rebel' versus 'rebel': see above).

(c) Fine-structure

We shall refer to variations of wave shape within

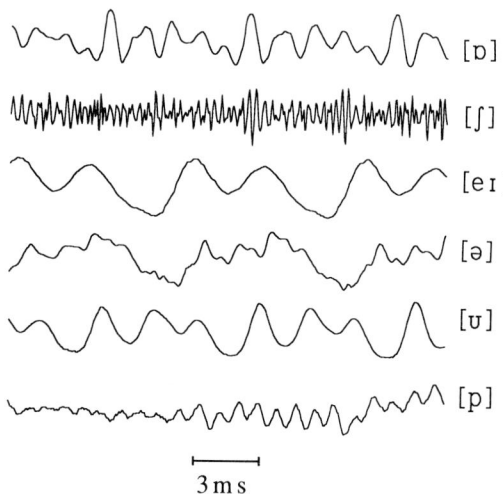

[ɒ]

[ʃ]

[eɪ]

[ə]

[ʊ]

[p]

3 ms

Figure 4. Sections of the speech pressure waveforms of figure 1, chosen so as to illustrate fine-structure information. Note the differences in wave shape among the four vowel-like sounds, and the generally higher fluctuation rate of /ʃ/ as compared to /p/, a reflection of the fact that /p/ has its spectral energy concentrated at lower frequencies than does /ʃ/.

single periods of periodic sounds, or over short time intervals of aperiodic ones as fine-structure information. This cue has dominant fluctuation rates from about 600 Hz to about 10 kHz. Acoustically, fine-structure informs about the spectrum of a sound (both amplitude and phase)‡ and so also contains its formant pattern. Fine-structure's primary auditory correlate relates to timbre, or quality, and it can convey at least two types of segmental linguistic information.

(i) Segmental cues to place of articulation and vowel quality

This potential function of fine-structure is by far the most important, not least because spectral shape variations are more or less the only acoustic cues to place. For example, it is well known that the two most important acoustic features that distinguish the words 'bait', 'date' and 'gate' from one another are the frequency spectrum of the initial release burst, and the dynamic formant transitions that follow (see, for example, Hazan & Rosen (1991) and references therein), all contained in fine-structure cues. Similarly, English voiceless fricatives in a prevocalic position may be distinguished from one another on the basis of static spectral shape, or the formant transitions in the following vowel, with the importance of each cue strongly dependent on the particular place of articulation (Harris 1958). Finally, spectral shape is the major cue to vowel identity (for examples, see figure 4).

‡ As no linguistic contrast depends primarily upon changes in phase spectrum, we discuss it no longer. It is important to realize, however, that the temporal structure of a speech signal (and especially its fine structure) will change with changes in the phase spectrum. So, the more a listener is relying on temporal cues for perception, the more important phase will be in determining auditory percepts (for a related point, see Rosen & Fourcin (1986)).

TEMPORAL FEATURES

		envelope	periodicity	fine-structure
segmental	manner	★	★	⋆
	voicing	⋆	★	★
	place			★
	voice quality	⋆		★
prosodic	tempo, rhythm	★		
	syllabicity	★		
	stress	⋆	★	
	intonation		★	

LINGUISTIC CONTRAST

Table 1. The potential role of various temporal features of speech in linguistic contrasts. The size of the stars indicates the extent to which a particular feature operates in a particular linguistic contrast, with a blank space indicating very weak or non-existent cues. Note that prosodic cues, which by definition occur across a number of segments, tend to be cued by the lower fluctuation rate categories. In fact, even for a linguistic feature such as intonation, related to periodicity, the relevant patterning (the fundamental frequency contour) occurs over much slower timescales than periodicity itself.

(ii) Segmental cues to voicing and manner

Voiced sounds have a spectrum heavily weighted to low frequencies (below about 1 kHz), and hence tend to have low fluctuation rates, whereas voiceless sounds typically have their peak energies at considerably higher frequencies, and thus tend to have high fluctuation rates (see, for example, the differences between the vowels and /ʃ/ in figure 4). First-formant transitions are known to play some role in distinguishing English voiced from voiceless plosives in initial prevocalic position (e.g. Soli 1983; Stevens & Klatt 1974). Apart from the manner information signalled by voicing, other cues to manner may be signalled by the shape of the spectrum. Nasals, for example, are characterized by a low first formant frequency, broad resonances, and zeros in the spectrum (Fujimura 1962). Stevens (1980, 1981) has discussed the role of sudden spectral changes (usually in conjunction with sudden envelope changes) in distinguishing consonantal sounds from non-consonantal ones.

3. CAVEATS

This list of correspondences between temporal features and linguistic information is not exhaustive, as there are weak potential cues to speech contrasts that are not mentioned above §. However, the most important

§ For example, fine structure may contain weak prosodic cues. At least in English, the vowels in unstressed syllables tend to be more neutral in quality than the vowels in stressed syllables. There is some evidence that this feature influences decisions of the 'rebel' versus 'rebel' variety (see above). So, too, may envelope give weak information about place of articulation, because the duration of aperiodicity in voiceless initial English plosives is known to vary lawfully with place of articulation (Lisker & Abramson 1964).

relationships are summarized in table 1. Note that temporal features may not always operate independently of one another. Duration, which has been grouped as an envelope cue, usually refers to the duration of an interval of speech with particular acoustic properties. For example, it is the duration of aperiodicity that helps to distinguish /tʃ/ from /ʃ/. Finally, there is likely to be much overlap in the frequency region over which the features operate. The release burst which is present in /tʃ/ but not in /ʃ/ is so short that its envelope certainly contains frequencies above 50 Hz. It is still the case, however, that the properties of the burst and affrication can be reasonably well divided among the features of envelope, periodicity and fine-structure.

4. APPLYING THE FRAMEWORK TO VARIOUS KINDS OF LISTENERS

When we come to consider the role of temporal features in speech perception, complications arise depending upon the type of listeners we are dealing with. For users of single-channel implants, no place–frequency mechanisms operate and so all auditory perception must be based on temporal features. Here, the three-way framework detailed above will apply completely, and the only extra problem concerns the

extent to which temporal features are modified by the patient's speech processor. Leaving this difficulty aside, it appears that most users of analogue single-channel implants can use envelope and periodicity information in a linguistic way, but the relative importance of each cue is not yet clear (Agelfors & Risberg 1989; Rosen & Ball 1986; Rosen *et al.* 1989; Tyler *et al.* 1987). In many, if not most, patients, there is also sensitivity to temporal fine-structure, but this is relatively rarely used in the perception of natural speech (Agelfors & Risberg 1987, 1989; Hochmair-Desoyer *et al.* 1985; Rosen & Ball 1986; Rosen *et al.* 1989; White 1983). However, the reception of unknown sentences (Hochmair-Desoyer *et al.* 1980, 1985) or accurate identification of place of articulation in consonants (Shannon *et al.* 1992) by auditory means alone implies some linguistic use of the temporal fine structure of speech.

At the other end of the observer continuum lie normal listeners, in whom the effects of peripheral auditory filtering must be considered‖. The normal auditory system decomposes a speech waveform, via the filtering action of the cochlea, into many waveforms, each of which will have its own three-way complement of temporal information (see figure 5). So, for instance, the envelope features transmitted by the auditory nerve will be modifications (to a greater

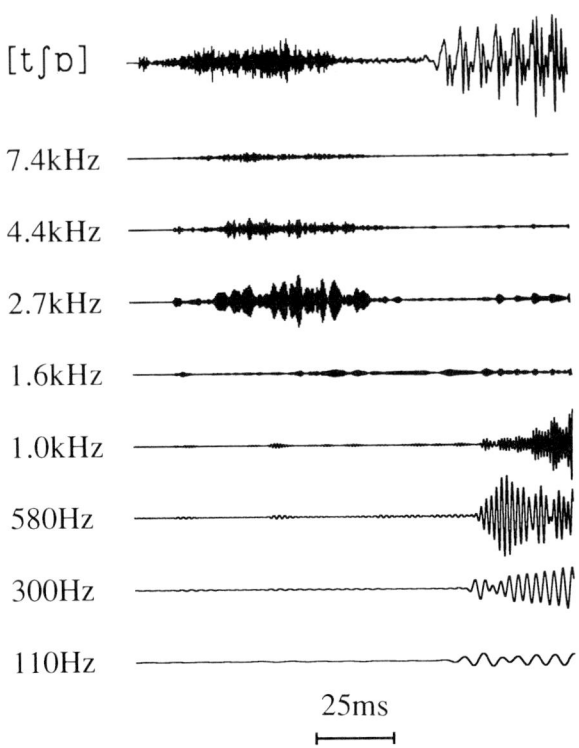

Figure 5. Original speech waveform of the initial part of the word 'chop', and waveforms resulting from auditory-like linear filtering at various centre frequencies. Note that the envelope features relating to the release burst, rise time and duration of the frication can only be found at the higher centre frequencies. At the same time, the periodicity of the vowel is only evident at lower centre frequencies.

‖ Plomp (1983) has described a three-way partition of the properties of speech sounds based on the concept of modulation which, although it does not purport to be a purely temporally-based description of speech, bears some resemblance to the system described here. In Plomp's words: 'Speech can be considered to be a wide-band complex signal modulated continuously in time in three different respects: (i) the vibration frequency of the vocal cords is modulated, determining the pitch variations of the voice; (ii) the temporal envelope of this signal is modulated by narrowing and widening the vocal tract locally by means of the tongue and lips; and (iii) the tongue and the lips in combination with the cavities of the vocal tract determine the sound spectrum of the speech signal, which may be considered as a modulation along the frequency scale.' These are clearly related to, although not identical with, the periodicity, envelope and fine-structure categories described above. There are, however, a number of difficulties in Plomp's characterization.

1. Fine structure is only discussed via the frequency domain (i.e. as a spectrum). Although perhaps reasonable for a system aimed solely at explaining the perception of normal listeners, it limits its usefulness as regards cochlear implants and theories of normal hearing which require use of temporal features in the perception of spectral shape variations.

2. No mechanism for distinguishing periodic from aperiodic sounds is proposed (the latter are never mentioned), nor is there any appropriate 'dimension' (i.e. frequency or time) suggested for the representation of periodicity (apart from its modulation in period over time). In fact, there appears to be the implication that all speech sounds are voiced.

3. Aspects of envelope are controlled by vocal fold behaviour, and not only by supralaryngeal manoeuvres (e.g. the decrease in amplitude over the vowel for any CV syllable uttered in citation form, as in 'key')

4. The spectrum of voiced speech sounds is influenced by the spectrum of the source (determined by vocal fold behaviour), as well as by the shape of the vocal tract. Generally speaking, it appears that a description in terms of acoustic properties is more useful than a description in terms of production, not least because it eliminates descriptive difficulties like the last two just mentioned. In terms of traditional descriptions of speech production, only the temporal feature of periodicity has a simple productive correlate: that of the source of excitation. Fine structure results, as implied above, from the interaction between source and filter, as indeed does envelope.

or lesser extent) of those observed on a speech waveform.

More importantly, peripheral auditory filtering means that temporal cues will be transformed into place cues, at least for periodicity and fine-structure. Although there is strong evidence that temporal properties of the signal have some role in the perception of periodicity and fine structure even in the normal listener, no one doubts that peripheral place–frequency analysis is crucial. Only for envelope features does it appear that temporal processes are nearly completely dominant¶.

Somewhere in the middle of this continuum on which listeners vary in the extent of their place–frequency analysis are hearing-impaired listeners and users of multi-channel implants. Moderately impaired listeners can retain a high degree of frequency selectivity, while those with more profound impairments can exhibit little or none (Faulkner *et al.* 1990). Multi-channel implant users, whose frequency selectivity is typically based on electronic filters in their speech processors, are probably comparable in selectivity to listeners who are severely or profoundly hearing impaired†. In any case, the greater the degradation of auditory frequency selectivity, the greater the role of explicitly temporal factors.

There is also a general limitation for any auditory system in the use of high fluctuation rates due to neurophysiological constraints on temporal coding. In particular, it is well known that mammalian auditory nerve fibres only synchronize their firing to acoustic sinusoids up to a maximum of about 5 kHz, so this must be an absolute upper limit on differential sensitivity to fine-structure and periodicity fluctuations (a limit that is probably also applicable to electrical stimulation).

5. FINAL REMARKS

There is still much to be learned about the role of the temporal structure of speech in influencing speech perception. Further clarification will lead not only to theoretical advances in understanding normal hearing, but also allow more efficient utilisation of limited auditory capacities in users of both acoustic and electro-cochlear auditory prostheses.

Thanks to Richard Baker, who assisted in making the figures, and Beverly Wright, Ken Robinson, Mike Johnson, Andy Faulkner, Bill Barry, Ginny Ball, Richard Baker, Michael Ashby and Evelyn Abberton for useful discussions or suggestions which greatly improved this paper. This work is supported by the Medical Research Council (U.K.).

¶ Even the auditory representation of envelope can be affected in minor ways by place–frequency analysis. For example, an aperiodic sound which is part of a burst of short duration will have a wider spectrum than the same sound turned on gradually.

† One important difference between implant users and profoundly hearing-impaired listeners is that the latter typically have a much more restricted range over which auditory stimulation is possible (perhaps only up to 1 kHz or so; Rosen *et al.* 1987). This kind of 'low-pass filtering' may have important implications for the representation of various temporal features (e.g. in periodicity versus aperiodicity, as most aperiodic energy is above 1 kHz).

REFERENCES

Agelfors, E. & Risberg, A. 1987 The identification of synthetic vowels by patients using a single-channel cochlear implant. *Proceedings of the XIth International Congress of Phonetic Sciences* **4**, pp. 181–184, Tallinn. (Also published in Speech Transmission Laboratory: Quarterly Progress and Status Report **2–3**, 31–38. Stockholm: Royal Institute of Technology. (1987))

Agelfors, E. & Risberg, A. 1989 Speech feature perception by patients using a single-channel Vienna 3M extracochlear implant. *Proceedings of the Speech Research '89 International Conference*, pp. 149–152. Budapest: Linguistics Institute of the Hungarian Academy of Sciences. (Also published in Speech Transmission Laboratory: Quarterly Progress and Status Report **1**, 145–149. Stockholm: Royal Institute of Technology. (1989))

Assmann, P.F. & Summerfield, Q. 1990 Modeling the perception of concurrent vowels: Vowels with different fundamental frequencies. *J. acoust. Soc. Am.* **88**, 680–697.

Bacon, S.P. & Viemeister, N.F. 1985 Temporal modulation transfer functions in normal-hearing and hearing-impaired listeners. *Audiology* **24**, 117–134.

Bailey, P.J. & Summerfield, Q. 1980 Information in speech: observations on the perception of [s]-stop clusters. *J. exp. Psychol.: Hum. Percept. Perform.* **6**, 536–563.

Burns, E.M. & Viemeister, N.F. 1976 Nonspectral pitch. *J. acoust. Soc. Am.* **60**, 863–869.

Crystal, D. 1969 *Prosodic systems and intonation in English*. Cambridge University Press.

Darwin, C.J. & Gardner, R.B. 1986 Mistuning a harmonic of a vowel: grouping and phase effects on vowel quality. *J. acoust. Soc. Am.* **79**, 838–845.

Dorman, M.F., Raphael, L.J. & Isenberg, D. 1980 Acoustic cues for a fricative-affricate contrast in word-final position. *J. Phonet.* **8**, 397–405.

Faulkner, A., Rosen, S. & Moore, B.C.J. 1990 Residual frequency selectivity in the profoundly hearing-impaired listener. *Br. J. Audiol.* **24**, 381–392.

Fry, D.B. 1968 Prosodic phenomena. In *Manual of phonetics* (ed. B. Malmberg), pp. 365–410. Amsterdam: North Holland.

Fujimura, O. 1962 Analysis of nasal consonants. *J. acoust. Soc. Am.* **34**, 1865–1875. (Also published in *Readings in acoustic phonetics* (I. Lehiste), pp. 238–248. Cambridge, Massachusetts: MIT Press. (1967))

Gerstman, L.J. 1957 Perceptual dimensions for the friction portions of certain speech sounds. Ph. D. thesis, New York University.

Harris, K.S. 1958 Cues for the discrimination of American English fricatives in spoken syllables. *Lang. Speech* **1**, 1–7. (Also published in *Acoustic phonetics* (ed. D. B. Fry), pp. 284–297. Cambridge University Press. (1976))

Hazan, V. & Rosen, S. 1991 Individual variability in the perception of cues to place contrasts in initial stops. *Percept. Psychophys.* **49**, 187–200.

Hochmair-Desoyer, I.J., Hochmair, E.S., Fischer, R.E. & Burian, K. 1980 Cochlear prostheses in use: recent speech comprehension results. *Arch. Oto-Rhino-Laryngol.* **229**, 81–98.

Hochmair-Desoyer, I.J., Hochmair, E.S. & Stiglbrunner, H.K. 1985 Psychoacoustic temporal processing and speech understanding in cochlear implant patients. In *Cochlear implants* (ed. R. A. Schindler & M. M. Merzenich), pp. 291–304. New York: Raven Press.

Howard, D.M. & Seligman, P.M. 1983 Initial comparisons between two simple time-domain fundamental frequency detectors. *Speech Hear. Lang.* (Prog. Rep. Dept. Phonetics & Linguistics, Univ. Coll. Lond.) **1**, 95–105.

Howell, P. & Rosen, S. 1983 Production and perception of rise time in the voiceless affricate/fricative distinction. *J. acoust. Soc. Am.* **73**, 976–984.

Howell, P. & Rosen, S. 1987 Perceptual integration of rise time and silence in affricate/fricative and pluck/bow continua. In *The psychophysics of speech perception* (ed. M. E. H. Schouten), pp. 173–180. Dordrecht: Martinus Nijhoff.

Lehiste, I. 1970 *Suprasegmentals*. Cambridge, Massachusetts: MIT Press.

Lisker, L. & Abramson, A.S. 1964 A cross-language study of voicing in initial stops: Acoustical measurements. *Word* **20**, 384–422.

Luxford, W.M., Berliner, K.I., Eisenberg, L.S. & House, W.F. 1987 Cochlear implants in children. *Annls Otol. Rhinol. Laryngol.* **96**(Suppl. 128), 136–138.

Mermelstein, P. 1975 Automatic segmentation of speech into syllabic units. *J. acoust. Soc. Am.* **58**, 880–883.

Miller, J.L. 1981 Effects of speaking rate on segmental distinctions. In *Perspectives on the study of speech* (ed. P. D. Eimas & J. L. Miller), pp. 39–74. Hillsdale, New Jersey: Lawrence Erlbaum.

Moore, B.C.J. & Glasberg, B.R. 1986 The role of frequency selectivity in the perception of loudness, pitch and time. In *Frequency selectivity in hearing* (ed. B. C. J. Moore), pp. 251–308. London: Academic Press.

Patterson, R.D. 1987 A pulse ribbon model of monaural phase perception. *J. acoust. Soc. Am.* **82**, 1560–1586.

Plomp, R. 1983 The role of modulation in hearing. In *Hearing – physiological bases and psychophysics* (ed. R. Klinke & R. Hartmann), pp. 270–276. Berlin: Springer-Verlag.

Repp, B.H., Liberman, A.M., Eccardt, T. & Pesetsky, D. 1978 Perceptual integration of acoustic cues for stop, fricative and affricate manner. *J. exp. Psychol.: Hum. Percept. Perform.* **4**, 621–637.

Rosen, S. 1986 Monaural phase sensitivity: frequency selectivity and temporal processes. In *Auditory frequency selectivity* (ed. B. C. J. Moore & R. D. Patterson), pp. 419–426. New York: Plenum.

Rosen, S. 1987 Phase and the hearing-impaired. In *The psychophysics of speech perception* (ed. M. E. H. Schouten), pp. 481–488. Dordrecht: Martinus Nijhoff.

Rosen, S. 1989 Temporal information in speech and its relevance for cochlear implants. In *Cochlear implant: acquisitions and controversies* (ed. B. Fraysse & N. Cochard), pp. 3–26. Basel: Cochlear AG.

Rosen, S. & Fourcin, A.J. 1986 Frequency selectivity and the perception of speech. In *Frequency selectivity in hearing* (ed. B. C. J. Moore), pp. 373–487. London: Academic Press.

Rosen, S. & Ball, V. 1986 Speech perception with the Vienna extra-cochlear single-channel implant: a comparison of two approaches to speech coding. *Br. J. Audiol.* **20**, 61–83.

Rosen, S., Walliker, J.R., Brimacombe, J.A. & Edgerton, B.E. 1989 Prosodic and segmental aspects of speech perception with the House/3M single-channel implant. *J. Speech Hear. Res.* **32**, 93–111.

Rosen, S., Walliker, J.R., Fourcin, A.J. & Ball, V. 1987 A microprocessor-based acoustic hearing aid for the profoundly impaired listener. *J. Rehab. Res. Develop.* **24**, 239–260.

Sachs, M.B. & Miller, M.I. 1985 Pitch coding in the auditory nerve: Possible mechanisms of pitch sensation with cochlear implants. In *Cochlear implants* (ed. R. A. Schindler & M. M. Merzenich), pp. 185–194. New York: Raven Press.

Sachs, M.B., Young, E.D. & Miller, M.I. 1983 Speech encoding in the auditory nerve: Implications for cochlear implants. In *Cochlear prostheses, an international symposium* (ed. by C. W. Parkins & S. W. Anderson) (*Ann. N.Y. Acad. Sci.* **405**), pp. 94–114. New York Academy of Sciences.

Seggie, D. 1986*a* The use of signal instantaneous frequency for voicing determination. *Speech Hear. Lang.* (Prog. Rep. Dept. Phonetics & Linguistics, Univ. Coll. Lond.) **2**, 179–192.

Seggie, D. 1986*b* The application of analytic signal analysis in speech processing. *Proceedings of the Institute of Acoustics Autumn Conference: Speech and Hearing, Windermere, U.K.* **8**, 85–92.

Shannon, R.V., Zeng, F.-G. & Wygonski, J. 1992 Speech recognition using only temporal cues. In *The auditory processing of speech: from sounds to words* (ed. M. E. H. Schouten). Berlin: Mouton De Gruyter. (In the press.)

Shinn, P. & Blumstein, S.E. 1984 On the role of the amplitude envelope for the perception of [b] and [w]. *J. acoust. Soc. Am.* **75**, 1243–1252.

Soli, S.D. 1983 The role of spectral cues in discrimination of voice onset time differences. *J. acoust. Soc. Am.* **73**, 2150–2165.

Stevens, K.N. 1980 Acoustic correlates of some phonetic categories. *J. acoust. Soc. Am.* **68**, 836–842.

Stevens, K.N. 1981 Constraints imposed by the auditory system on the properties used to classify speech sounds: data from phonology, acoustics and psycho-acoustics. In *The cognitive representation of speech* (ed. T. F. Myers, J. Laver & J. Anderson), pp. 61–74. Amsterdam: North Holland.

Stevens, K.N. & Blumstein, S.E. 1981 The search for invariant acoustic correlates of phonetic features. In *Perspectives on the study of speech* (ed. P. D. Eimas & J. L. Miller), pp. 1–38. Hillsdale, New Jersey: Lawrence Erlbaum.

Stevens, K.N. & Klatt, D.H. 1974 Role of formant transitions in the voiced-voiceless distinction for stops. *J. acoust. Soc. Am.* **55**, 653–659.

Summerfield, Q., Bailey, P.J., Seton, J. & Dorman, M.F. 1981 Fricative envelope parameters and silent intervals in distinguishing 'slit' and 'split'. *Phonetica* **38**, 181–192.

Tyler, R.S., Tye-Murray, N., Preece, J.P., Gantz, B.J. & McCabe, B.F. 1987 Vowel and consonant confusions among cochlear implant patients: do different implants make a difference? *Annls Otol. Rhinol. Laryngol.* **96**(Suppl. 128), 141–144.

Umeda, N. 1975 Vowel duration in American English *J. acoust. Soc. Am.* **58**, 434–445.

Van Tasell, D.J., Soli, S.D., Kirby, V.M. & Widin, G.P. 1987 Speech waveform envelope cues for consonant recognition. *J. acoust. Soc. Am.* **82**, 1152–1161.

White, M.W. 1983 Formant frequency discrimination and recognition in subjects implanted with intracochlear stimulating electrodes. In *Cochlear prostheses, an international symposium* (ed. C.W. Parkins & S.W. Anderson) (*Ann. N.Y. Acad. Sci.* **405**), pp. 348–359. New York Academy of Sciences.

Pitch related to spectral edges of broadband signals

ARMIN KOHLRAUSCH AND ADRIANUS J. M. HOUTSMA

Institute for Perception Research (IPO), P.O. Box 513, 5600 MB Eindhoven, The Netherlands

SUMMARY

A complex tone often evokes a pitch sensation associated with its extreme spectral components, besides the holistic pitch associated with its fundamental frequency. We studied the edge pitch created at the upper spectral edge of complexes with a low-pass spectrum by asking subjects to adjust the frequency of a sinusoidal comparison tone to the perceived pitch. Measurements were performed for different values of the fundamental frequency and of the upper frequency of the complex as well as for three different phase relations of the harmonic components. For a wide range of these parameters the subjects could adjust the comparison tone with a high accuracy, measured as the standard deviation of repeated adjustments, to a frequency close to the nominal edge frequency. The detailed dependence of the matching accuracy on temporal parameters of the harmonic complexes suggests that the perception of the edge pitch in harmonic signals is related to the temporal resolution of the hearing system. This resolution depends primarily on the time constants of basilar-membrane filters and on additional limitations due to neuronal processes.

1. INTRODUCTION

Broadband signals with a steep transition in the amplitude spectrum can produce pitch sensations related to the frequency of the spectral edge. These edge pitches have been mostly studied for band-limited noise signals (see, for example, Small & Daniloff (1967); Rakowski (1968); Fastl (1971, 1980); Klein & Hartmann (1981)). Fastl (1980), for instance, measured the pitch strength of low-pass noise as a function of the spectral slope at the cut-off frequency. For an edge frequency of 1 kHz, he found that the pitch strength decreases as the spectral slope was reduced below -36 dB per octave. Even for very steep filter slopes, however, the pitch strength associated with noise edges is much lower (more than a factor five) than the pitch strength of a pure sinusoid (Fastl & Stoll 1979).

It has been proposed that the noise edge pitch is due to lateral inhibition in the hearing system (v. Bekesy 1960, 1963; Small & Daniloff 1967). As in the case of optical 'Mach bands', the excitation caused by the stimulus is raised at the spectral edge. This relative maximum in excitation then leads to a pitch sensation.

In the present article, we study pitch effects created at spectral edges of harmonic complex sounds. The main difference between these stimuli and band-limited noise is found in the temporal waveform. Besides periodicity, the waveforms show high-frequency ripples in their fine structure which closely correspond to the spectral-edge frequency. These pure-tone-like parts of the waveform alternate with peaks, and their relative duration within each period depends on the harmonic number of the edge component. This relation is illustrated by the two signals in

figure 1, which both have an edge frequency of 2 kHz. The left signal has a fundamental frequency of 100 Hz and the highest harmonic has order 20, whereas for the other signal, with a fundamental of 250 Hz, the order is only 8. Both complexes have a flat spectrum and the components are added in zero phase.

An alternative way to emphasize the spectro-temporal properties of complex tones is a short-time Fourier analysis. Figure 2a, b shows such an analysis for the two sounds from figure 1. The short-time spectra in both panels are calculated using a Hanning window of 10 ms duration. The time axis, running from front to back, covers approximately 25 ms. The temporal shift between adjacent spectra is 0.5 ms. Figure 2a shows the complex with 100 Hz fundamental and reveals two different spectral patterns alternating in time. One is a flat spectrum showing the spectral extent of the complex from 100 to 2000 Hz. This pattern results from those time points where the window is centred on a pulse of the temporal waveform. The other pattern (visible, for example, in the first spectrum) shows distinct maxima at the edge frequencies, slightly amplitude-modulated in time. This pattern results from window positions in between the peaks.

Such an alternating pattern is only seen if the window covers no more than one period of the sounds. In the left-hand panel, the total window length is exactly equal to one period. The effective window duration, however, is shorter due to the raised-cosine ramps of the Hanning window. In figure 2b the same calculation is performed for the complex with 250 Hz fundamental. In this case, the window length is 2.5 times the period of the time signal and therefore, the short-time spectra emphasize the individual harmon-

Phil. Trans. R. Soc. Lond. B (1992) **336**, 375–382

Printed in Great Britain

375

© 1992 The Royal Society

[81]

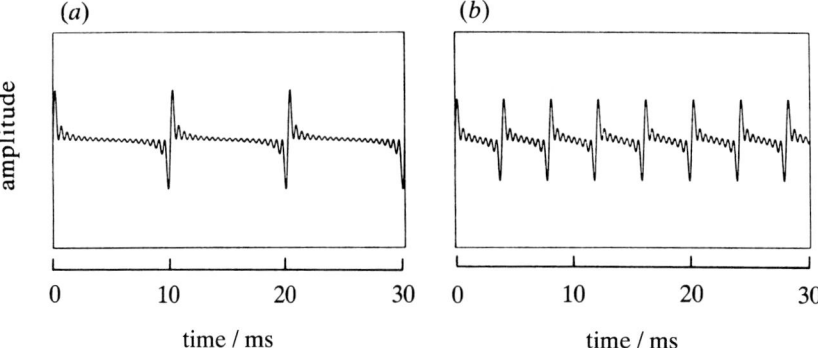

Figure 1. Time functions of two harmonic complexes with an upper-edge frequency of 2 kHz: (*a*) shows a complex with 100 Hz fundamental frequency, consisting of 20 harmonics, (*b*) shows a complex with 250 Hz fundamental and 8 harmonics. All components in the complexes have the same amplitude and zero starting phase.

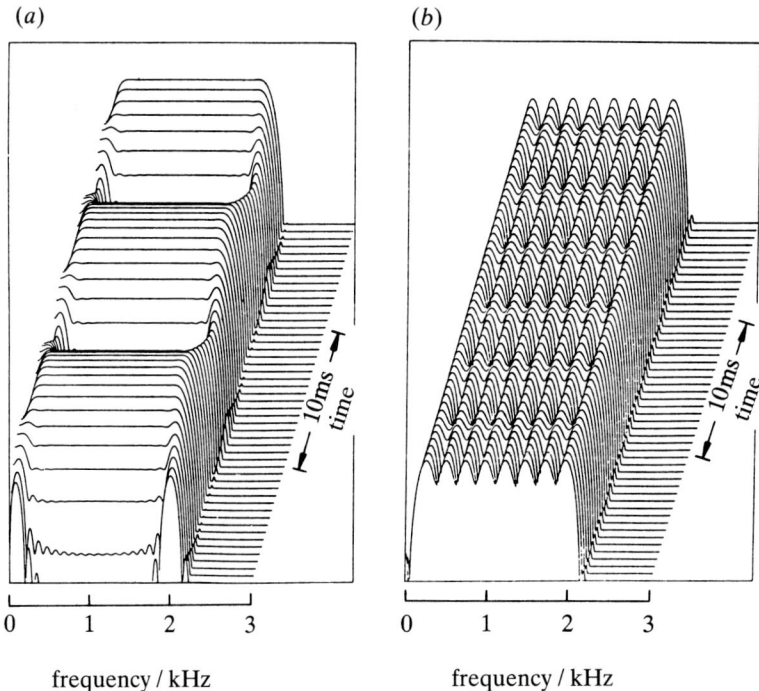

Figure 2. Short-time spectra of the two sounds from figure 1. The spectra are calculated using a Hanning window of 10 ms duration.

ics of the complex. In other words, the creation of a spectral peak at the edge frequency as shown in the left-hand panel of figure 2 will depend on the effective time constants of the analysing system.

Edge-pitch effects of harmonic complexes have been mentioned previously in the framework of complex sound perception by several authors (Martens 1981, Mehrgardt 1982; Patterson 1987; Kohlrausch 1988; Moore & Glasberg 1989). Martens mentioned that for low-pass complexes with a flat spectrum up to 4 kHz, a faint high-frequency tone was perceived, if F_0 did not exceed 100 Hz. He compares this effect with the 'tonal quality of low-pass noise' (Martens 1981, p. 235). Moore and Glasberg, on the other hand, remarked that band-limited harmonic complexes create a clear, sine-tone-like pitch, which is perceptually quite different from the pitch created by noise

bands. They also provided individual frequency matches to a complex tone with a fundamental frequency of 50 Hz and an upper edge at 1000 Hz. The comparison tone was adjusted on average by the three subjects to frequencies 15 to 75 Hz above the nominal edge. For a complex with F_0 equal to 100 Hz and an upper edge at 2 kHz, the average matched frequency was 2.054 kHz.

There are thus some hints in the literature that edge pitches of periodic sounds have sources other than or additional to edge effects of noise bands. The present article investigates how accurately the pitch at the upper spectral edge of harmonic complexes can be matched with a sinusoidal comparison tone. This quantitative measure is then used to determine the existence region of the edge pitch as well as the accuracy of the matches for a wide range of the following parameters: fundamental and upper-edge frequency of the complex, order of the highest component and phase values of the individual components of the complex.

2. METHOD

The test sounds consisted of equal-amplitude sinusoids with a common fundamental frequency F_0. All components below the upper-edge frequency were present in the complex. The stimuli had a duration of 500 ms and were shaped with 20 ms raised-cosine ramps. They were presented diotically to the subjects via headphones (TDH 49) at an average sound pressure level of 65 dB. Sounds were generated digitally and converted by either a 12-bit or a 16-bit D/A converter at a sampling rate of 10 kHz.

Sinusoidal comparison tones had the same duration (500 ms) and alternated with the test sounds with a 500 ms silent interval in between. The frequency of the comparison tone was adjusted by the subjects with an unmarked ten-turn potentiometer, which controlled the frequency of an oscillator (Philips P5190). The potentiometer received a random offset at the beginning of each new measurement in order to avoid systematic errors. There was no temporal limit for the matching procedure, but a typical match took about 30 s. The finally adjusted frequency was taken as the data point for further evaluation.

Three subjects, among whom were the two authors, participated in the experiments and performed ten matches for each test complex. Within one experimental session, typically seven different complexes were matched five times. In successive matches, different complexes were presented. From the ten matches, the mean and unbiased standard deviation were calculated. Evident mismatches (e.g. octave confusions) were omitted.

3. RESULTS

In a pilot experiment the distribution of the matched frequencies for one typical complex and a larger number of matches was examined. The test stimulus was a complex with $F_0 = 100$ Hz, an upper-edge frequency of 2 kHz and zero phases of the components (cf. figure 1a). In figure 3, the distribution of matched frequencies is shown separately for the three subjects. All subjects were able to adjust the frequency of the comparison tone within a narrow frequency range of approximately 50 Hz width. The standard deviation is very similar for the three listeners and lies between 11.2 and 12.7 Hz. This corresponds to a relative accuracy of about 0.6%. The means of the matches show a greater variation and lie between 2019 and 2072 Hz. It turned out in the measurements that the shift between nominal and average matched edge frequency depended very much on the individual listener, while the accuracy (spread) of the matches was more similar. On the basis of this initial observation we decided to use the standard deviation of repeated matches as an indication of how clearly the edge pitch could be perceived by the subjects. The data points in the following figures are calculated by

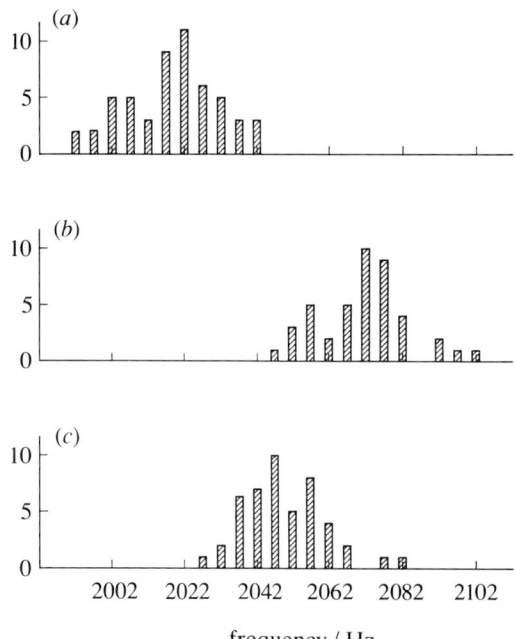

Figure 3. Distribution of matched frequencies for the signal in figure 1a with a nominal edge at 2 kHz. The abscissa denotes the frequency of the matches, the ordinate gives the number of matches within a bin of 5 Hz. Individual results for three listeners: (a) A.K., $n = 54$, $\bar{x} = 2019.1$ (12.7) Hz; (b) A.H., $n = 43$, $\bar{x} = 2071.6$ (11.9) Hz; (c) N.V., $n = 46$, $\bar{x} = 2049.3$ (11.2) Hz; n is total number of matches for this sound.

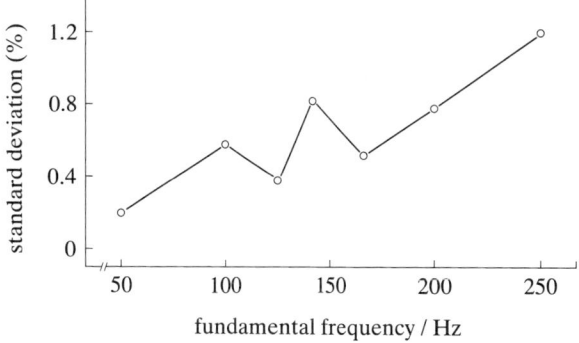

Figure 4. Standard deviation in percent of the edge frequency for complexes with a constant upper-edge frequency of 2 kHz and different fundamental frequencies. Averages of the three listeners.

averaging the individual spreads of the three observers.

In the first experiment, a set of test sounds was used having a constant upper-edge frequency of 2 kHz but different values of the fundamental frequency. The fundamental frequency varied between 50 and 250 Hz. Therefore, the order of the highest component decreased from 40 to 8 and the duration of a period decreased from 20 ms to 4 ms. In figure 4, the standard deviation of the matches (as a percentage of the edge frequency) is plotted as a function of the fundamental frequency. For all values of F_0, the subjects could clearly hear a pitch related to the edge frequency. The accuracy of the matches, however, decreases with increasing fundamental frequency. The average accuracy for $F_0 = 50$ Hz is about 0.2% and the value at 250 Hz fundamental is about 1.2%. In a purely spectral (or place) view, this increase could be a consequence of the decreasing density of spectral components for higher values of F_0. This, however, contradicts the observation with low-pass-filtered noise signals which have the highest spectral density. Typical accuracies for edge pitches of such signals are in the order of several percent (Fastl 1971; Klein & Hartmann 1981). In a temporal view, one has to consider that the increase in F_0 leads to a decrease in the period of the waveform. Therefore, if the waveform is analysed with a fixed temporal window, more envelope peaks fall into one window and the high-frequency ripple becomes less prominent.

In the following experiment, the influence of the fundamental frequency was studied for a wider range of edge frequencies. Complexes of three fundamental frequencies were used, namely 50 Hz, 100 Hz, and 200 Hz. The upper-edge frequency was increased in steps of 200 Hz from 600 Hz (800 Hz for 200 Hz fundamental) to 2000 Hz. The results in figure 5 are again plotted as averages over the three observers. The data for 50 Hz are represented by circles, the 100 Hz data by squares and the 200 Hz data by triangles. At all edge frequencies, the 50 Hz complex leads to the lowest standard deviation. The values for these complexes decrease towards higher edge frequencies from about 0.4% to 0.2%. The order of the highest harmonic varies between 12 and 40 and we can thus assume that the edge component is in all cases not resolved.

This is obviously not the case for the two other fundamental frequencies. Thus, for these two data sets the edge frequencies cover a transition region from well resolved (at the left-hand side) to less or not resolved harmonics (at the right-hand side). The accuracy of the match always remains less than for the 50 Hz complex and at frequencies above 1 kHz, the difference is a factor of two or more. The fact that the 50 Hz complex with a period of 20 ms gives rise to a much clearer edge pitch than complexes with a higher fundamental and thus a shorter period, supports the finding from the previous experiment.

For a comparison with the above-mentioned results from Moore & Glasberg (1989), we should also mention the average shift for the 50 Hz complex with an edge frequency of 1000 Hz. The average matched frequency for the three observers lies 17 to 35 Hz above 1000 Hz. This result is well within the range found by Moore & Glasberg.

As mentioned previously, one important difference between edge pitches of complex tones and of low-pass noises lies in the accuracy of matches to the pitch. If, on the other hand, the edge pitch has a quality very similar to a pure sinusoid – especially at low fundamental frequencies – one might expect the accuracy of matches to come very close to that of pure tones. Unfortunately, literature data for pure-tone matches show a great variety, depending on the specific procedure and subject's training. For the method of adjustment and highly trained subjects, an accuracy of 0.02% to 0.08% has been reportd (Nordmark 1968; Rakowski 1971).

To have a reliable comparison with the complex-tone matches, we performed matches to pure tones and to low-pass noises with the same apparatus and subjects as used in the complex-tone measurements. The sinusoidal stimuli were generated digitally and D/A converted at a rate of 20 kHz. The noise stimuli were generated by an analogue noise generator followed by a digitally adjustable low-pass filter with a spectral slope of 180 dB per octave. These measurements were performed for (edge) frequencies between 1000 and 4000 Hz. In the lower octave, matches were obtained every 200 Hz. In the higher octave, matches were obtained every 400 Hz.

Figure 6 shows the results of these two measurements. Open symbols indicate matches to pure tones and the results must be referred to the left-hand scale. Noise matches are represented by the filled symbols and are referred to the right-hand axis, which has a scale factor of 10. The spread of the pure-tone matches is around 0.1% at 1 kHz, increases slightly above 2 kHz and reaches a value of 0.25% at the highest tested frequency of 4 kHz. Such an increase of the relative accuracy has been reported previously for frequency difference limens (Moore 1974).

As expected from literature data, the noise matches

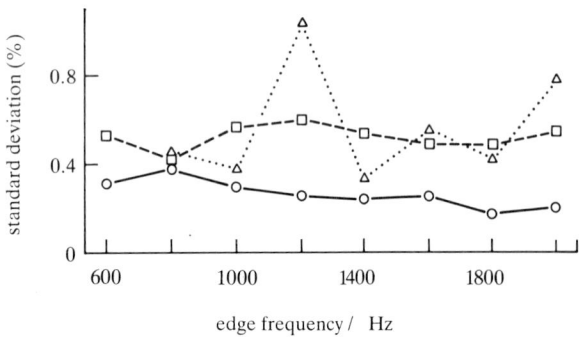

Figure 5. Standard deviation in percent of the edge frequency for complexes with a fundamental frequency of 50 Hz (circles), 100 Hz (squares) and 200 Hz (triangles).

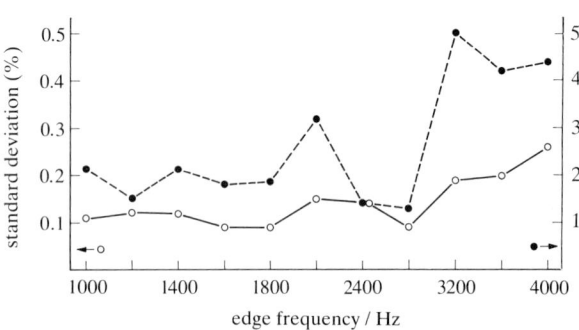

Figure 6. Standard deviation of the matches to sinusoids (open symbols) and to low-pass-filtered noise bands (filled symbols) as a function of the (edge) frequency. The left-hand ordinate has to be used for the pure-tone data and the right-hand ordinate has to be used for the noise data.

are an order of magnitude less accurate. In the low-frequency region, the average accuracy is around 2% and this value increases to 5% at the highest frequency tested. A possible restriction for the noise data could lie in the limited slope at the spectral transition. Although comparable measurements with digitally generated frozen noise (additive synthesis) yielded some improvement over the running-noise conditions, standard deviations were still much larger than those obtained with tone complexes.

If we compare these accuracies with the results from the complex-tone matches, we can conclude that the matches to the 50 Hz complexes are less than a factor of two worse in terms of accuracy than the pure-tone matches. Also, the standard deviations observed for higher fundamental frequencies are still significantly lower than typical values for noise edges. This can be taken as an indication that temporal properties of the complex sounds play an important role in the perception of the edge tones.

To test further the temporal aspects of the edge pitch, in the last experiment we varied the phase values in the complex. A strong phase effect, with an otherwise constant spectrum, indicates the role of temporal effects. The complexes in this measurement had a fundamental frequency of 100 Hz and an upper edge between 600 and 2000 Hz, which was increased in steps of 200 Hz. The choice of phase values was inspired by earlier results from masking experiments, from which conclusions about the internal excitation of the complexes were drawn (Smith *et al.* 1986; Kohlrausch 1988). Besides a zero-phase complex, two complexes with a positive or a negative Schroeder phase were used. The formula for the Schroeder-phase values for a flat-spectrum complex is as follows (Schroeder 1970):

$$\phi_n = \pm \pi * n(n-1)/N. \qquad (1)$$

In this formula, n indicates the order of the individual harmonic and N gives the total number of harmonics in the complex. The resulting temporal waveform for both signs of equation (1) has a relatively flat envelope. However, after transformation in the hearing periphery, the waveform of the '+ complex' is transformed into a very peaked waveform which can be even more peaked than the internal waveform of the sine-phase complex (Kohlrausch 1988). One can therefore expect that for this complex the edge pitch is as clearly audible as for the zero-phase complex. The '− complex', on the other hand, has a very flat envelope even after cochlear filtering. Therefore, its edge pitch should be less pronounced.

In Figure 7, we present the results for these three complexes. The different symbols represent the three phase relations of the masker: circles indicate zero phase, open triangles the positive Schroeder phase and filled triangles the negative Schroeder phase. As expected from the results of masking experiments, the positive Schroeder phase leads to results very similar to the zero-phase complex. This similarity is not only observed for the accuracy, but holds also for the subject-dependent shift between nominal and matched frequency (Kohlrausch & Houtsma 1991).

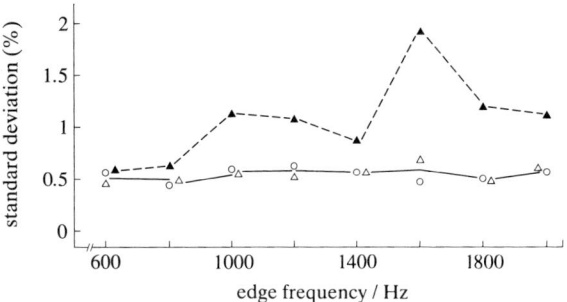

Figure 7. Standard deviation of the matches to complexes with a fundamental frequency of 100 Hz and various upper-edge frequencies. The three symbols indicate different phase choices for the components. Circles: zero phase; open triangles: positive Schroeder phase; filled triangles: negative Schroeder phase. Averages of the three listeners.

Comparing the three curves at different frequencies, the increasing influence of phase effects at higher frequencies becomes obvious. Here, the filled symbols are about a factor two above the open symbols. At the left-most data point, however, the two sets of curves approach each other. The data point at the lowest edge frequency corresponds to an order of six for the edge component. Components of such a low order are generally assumed to be well resolved in the hearing system (Plomp 1976). This means that the interaction between harmonics is minimal and therefore the relative phase between adjacent components should not influence the internal representation (Goldstein 1967).

In summary, the following results were obtained in the pitch-matching experiments: The pitch related to the upper spectral edge in harmonic complex tones can be matched quite well with a sinusoidal comparison tone. The standard deviation calculated from repeated matches appears to be very similar for the three subjects, whereas the average matched frequency differs significantly between subjects (figure 3). For a fixed edge frequency, complexes with a lower value for F_0 and thus a higher harmonic order of the edge component create a clearer edge pitch (figures 4 and 5). Matches to complex-tone edges are much more accurate than matches to noise edges. On the other hand, matches to a pure tone are only a factor two more accurate than matches to a complex with F_0 equal to 50 Hz (figure 6). The accuracy of the matches depends on the phases of the masker components. This phase effect only disappears, if the edge component has a low harmonic number of 6 or 8 (figure 7).

4. DISCUSSION

In this section, we will try to relate the experimental observations to known properties of the hearing system. First of all, the results indicate a large and possibly fundamental difference between the upper edge pitch of harmonic complex sounds and of low-pass-filtered noise. While the noise edge pitch can be matched with an accuracy of several percent, the complex-tone edge pitch, at least for low fundamental frequencies and a high order of the highest harmonic, can be matched with an accuracy well below one percent. This accuracy is similar to values obtained in tone-on-tone matching (see Letowski 1982) and it supports the subjective impression of a pure-tone-like pitch percept, which is superimposed on the virtual pitch related to the fundamental frequency of the complex. As was already mentioned in connection with the time-function plot of figure 1, the broadband waveform contains the edge frequency in the fine structure. It remains to be checked whether this sinusoidal waveform will be present after cochlear filtering.

Because such an analysis has been performed previously by Duifhuis (1970) in the context of a related experiment, we will first recall some of his findings. Duifhuis investigated the percept of periodic pulse sequences with fundamental frequencies 25, 50 and 100 Hz. The spectrum of such a sequence is a series of cosines with constant amplitude and zero starting phase. The task of the subjects was to vary the amplitude of a certain harmonic until the harmonic was just audible as a pure tone, characterized by a given pitch. This paradigm can be performed for every harmonic by increasing its amplitude. What is particularly relevant in the context of the present discussion is the observation that for high-order harmonics (above an order of 16), also a decrease in the amplitude leads to a pitch percept. The decrease in the amplitude is equivalent to the addition of a cosine in antiphase with an amplitude lower than the amplitude of the individual components in the complex. Thus, the waveform of a pulse sequence with a suppressed harmonic is very similar to the waveform of a low-pass pulse sequence with its cutoff at the suppressed harmonic. Close inspection shows that the amplitude of the high-frequency ripple in a low-pass complex is half the amplitude of the ripple in a complex with a completely suppressed harmonic, given that the individual harmonics have the same amplitude in both signals.

Duifhuis explained his finding on the basis of waveforms resulting from a band-pass filter tuned to the suppressed harmonic. For a sufficiently high quality factor of the filter (Duifhuis proposes a value of about 10, a value in the range of recent auditory-filter estimates according to Moore & Glasberg (1983)) the suppressed harmonic is seen within each period of the filtered waveform as a pure tone. He further assumes that an appropriate detector which is only active during that part of the period showing this tone-like property could respond to this sinusoidal waveform.

For the explanation of our result we can follow this reasoning very closely, as did Moore & Glasberg (1989) in their analysis of the edge tone. That is, for high harmonic numbers of the edge component and zero phases of all components, the fine structure of the edge tone will be preserved in each period of the signal resulting from the inner-ear filter tuned to the edge frequency. The higher the harmonic number, the greater the part within each period showing this pure-tone-like fine-structure. The increasing accuracy of the matches for increasing harmonic number of the highest component (figure 4) indicates that this temporal information is indeed used by the subjects.

The observation of a phase effect further supports this explanation. Randomizing of the phase values reduces the peak factor of the broadband signal, and also of the filtered waveforms. In such a stimulus, there are no longer time sections within a period where high excitation in the edge-frequency region is contrasted with low excitation in lower spectral regions. Thus, the temporal fine-structure cue is no longer available. The increase in the standard deviation for the negative Schroeder phase for higher harmonic numbers (figure 7) supports this view, but the results do also indicate that additional, probably purely spectral cues, still allow for a pitch percept for this complex.

While Duifhuis (1970) and Moore & Glasberg (1989) only considered the consequences of band-pass filtering, we have some indications that basilar-membrane properties are not the only critical parameters for the perception of the edge pitch. This can be demonstrated by measuring edge pitches for complexes always having the same order of the highest component, e.g. 10 or 14. Following the above arguments, one would expect the same accuracy of matches for all fundamental frequencies, since the relative bandwidth of the basilar-membrane filters is, as a first approximation, independent of frequency above 500 Hz. The accuracy of the pitch matches, however, decreases significantly for fundamental frequencies above about 250 to 300 Hz, where the edge frequencies are in the region 3 to 4 kHz (Kohlrausch & Houtsma 1991).

One can think of two possible causes of this effect. On the one hand, the edge pitch seems to disappear for edge frequencies of 3 kHz and higher, i.e. the breakdown occurred at a lower F_0 with the edge at harmonic order 14 than with the edge at harmonic order 10. Consequently, the absolute frequency of the breakdown appears to remain fairly constant, which could be due to a decrease in neural phase locking. In the auditory nerve such a decrease begins around 1 kHz (e.g. Rose *et al.* 1967), and can be modelled by a half-wave rectifier and a subsequent low-pass filter with cutoff frequency 1 kHz following each auditory filter. A loss in phase locking should make it more difficult for a subsequent detector to recognize the edge tone in the complex pattern resulting from cochlear filtering.

From the viewpoint of temporal resolution, on the other hand, one has to consider the decreasing duration of a period for increasing F_0. A fundamental frequency of 300 Hz has a period of 3.3 ms and such a duration agrees with the estimates of the minimal integration time of the hearing system (e.g. Green 1973). Following this consideration, the period of the complex would become so short that the envelope modulation, which is clearly present at the level of the basilar membrane, would no longer be represented in higher stages of the hearing pathway.

A similar argument is based on the effects of

adaptation and forward masking, as they are seen for instance in non-simultaneous physiological masking experiments (Harris & Dallos 1979). If we apply the arguments presented in Kohlrausch *et al.* (1992), each amplitude peak in a waveform suppresses immediately following signal parts of lower amplitude. Because this temporal effect is thought to be a consequence of neural adaption, its duration should be independent of frequency. Such a temporal suppression will not be critical for the perception of the edge pitch, as long as the period of the stimulus is sufficiently long. However, for shorter periods, the suppression will affect the fine-structure information relevant for the pitch percept. Such an explanation makes it reasonable that an F_0 value of 50 Hz leads to much smaller standard deviations than F_0 values of 100 or 200 Hz (cf. figure 5).

In summary, all the data presented in this article are in line with previously published experimental data and models about how non-stationary acoustic signals are represented in the hearing pathway. On the other hand, it seems quite difficult to deduce precise quantitative predictions from the data about, for example, the ringing time of the peripheral filters or the frequency dependence of phase locking. This is due to the fact that a global measure such as pitch perception accuracy depends on many more stages of the hearing pathway than peripheral ones. What we definitely can say is that, from all we know, the information necessary for a pitch percept is retained in the peripheral stages of the hearing system.

REFERENCES

Bekesy, G.v. 1960 *Experiments in hearing*. New York: McGraw-Hill.

Bekesy, G.v. 1963 Hearing theories and complex sounds. *J. acoust. Soc. Am.* **35**, 588–601.

Duifhuis, H. 1970 Audibility of high harmonics in a periodic pulse. *J. acoust. Soc. Am.* **48**, 888–893.

Fastl, H. 1971 Über Tonhöhenempfindungen bei Rauschen. *Acustica* **25**, 350–354.

Fastl, H. 1980 Pitch strength and masking patterns of low-pass noise. In *Psychophysical, physiological, and behavioral studies in hearing* (ed. G. v.d. Brink & F. Bilsen), pp. 334–339. Delft University Press.

Fastl, H. & Stoll, G. 1979 Scaling of pitch strength. *Hear. Res.* **1**, 293–301.

Goldstein, J.L. 1967 Auditory spectral filtering and monaural phase perception. *J. acoust. Soc. Am.* **41**, 458–479.

Green, D.M. 1973 Minimum integration time. In *Basic mechanisms in hearing* (ed. A. R. Moeller), pp. 829–846. New York: Academic Press.

Harris, D.M. & Dallos, P. 1979 Forward masking of auditory nerve fiber responses. *J. Neurophysiol.* **42**, 1083–1107.

Klein, M.A. & Hartmann, W.M. 1981 Binaural edge pitch. *J. acoust. Soc. Am.* **70**, 51–61.

Kohlrausch, A. 1988 Masking patterns of harmonic complex tone maskers and the role of the inner ear transfer function. In *Basic issues in hearing* (ed. H. Duifhuis, J. W. Horst & H. Wit), pp. 339–350. London: Academic Press.

Kohlrausch, A. & Houtsma, A.J.M. 1991 Edge pitch of harmonic complex tones. *IPO A. Prog. Rep.* **26**, 39–49.

Kohlrausch, A., Püschel, D. & Alphei, H. 1992 Temporal

resolution and modulation analysis in models of the auditory system. In *The auditory processing of speech: from sounds to words* (ed. M. E. H. Schouten). Berlin: Mouton-De Gruyter. (In the press.)

Letowski, T. 1982 A note on the difference limen for frequency differentiation. *J. Sound Vibr.* **85**, 579–583.

Martens, J.-P. 1981 Audibility of harmonics in a periodic complex. *J. acoust. Soc. Am.* **70**, 234–237.

Mehrgardt, S. 1982 Psychoakustische Untersuchungen an harmonischen Tonkomplexmaskierern. Ph. D. thesis, University of Göttingen.

Moore, B.C.J. 1974 Relation between the critical bandwidth and the frequency difference limen. *J. acoust. Soc. Am.* **55**, 359.

Moore, B.C.J. & Glasberg, B. 1983 Suggested formulae for calculating auditory-filter bandwidths and excitation patterns. *J. acoust. Soc. Am.* **74**, 750–753.

Moore, B.C.J. & Glasberg, B. 1989 Difference limen for phase in normal and hearing-impaired subjects. *J. acoust. Soc. Am.* **86**, 1351–1365.

Nordmark, J.O. 1968 Mechanisms of frequency discrimination. *J. acoust. Soc. Am.* **44**, 1533–1540.

Patterson, R.D. 1987 A pulse ribbon model of monaural phase perception. *J. acoust. Soc. Am.* **82**, 1560–1586.

Plomp, R. 1976 *Aspects of tone sensation—A psychophysical study*. London: Academic Press.

Rakowski, A. 1968 Pitch of filtered noise. Proceedings of the 6th International Congress on Acoustics, A-105.

Rose, J.E., Brugge, J.F., Anderson, D.J. & Hind, J.E. 1967 Phase-locked response to low-frequency tones in single auditory nerve fibers of the squirrel monkey. *J. Neurophysiol.* **30**, 769–793.

Schroeder, M.R. 1970 Synthesis of low-peak-factor signals and binary sequences with low autocorrelation. *IEEE Trans. Inf. Theory* **16**, 85–89.

Small, A.M.Jr & Daniloff, R.G. 1967 Pitch of noise bands. *J. acoust. Soc. Am.* **41**, 506–512.

Smith, B.K., Sieben, U.K., Kohlrausch, A. & Schroeder, M.R. 1986 Phase effects in masking related to dispersion in the inner ear. *J. acoust. Soc. Am.* **80**, 1631–1637.

Discussion

E. F. Evans (*Department of Communication and Neuroscience, University of Keele, U.K.*). Has Professor Houtsma investigated the effects of stimulus level?

From experiments looking at cochlear nerve fibre temporal discharge patterns to stimuli as those producing the 'rabbit ear' spectra, major differences are found in response at low stimulus levels (between 20 dB of threshold) and high levels (about 50 dB above threshold), such that the psychophysical effects that you are detecting should be those corresponding to the physiological responses obtained at the higher stimulus levels. In the physiological case, the systematic differences with level tend to disappear under two conditions: (i) increasing the fundamental from 40 to 125 Hz; (ii) randomizing the phase, both very much in line with Professor Houtsma's psychophysical findings.

Another prediction from the physiological data is that the pitch of the upper edge of the band of harmonics should be expected to increase with stimulus level from within the band at low levels to slightly outside at high level.

All these results can be modelled by a linear filter followed by an automatic gain control of about 10 ms time constant in a simple cochlear nerve model (Evans 1980, 1987, 1988).

References

Evans, E.F. 1980 An electronic analogue of single unit recording from the cochlear nerve for teaching and research. *J. Physiol., Lond.* **298**, 6–7.

Evans, E.F. 1987 Modelling cochlear nerve fibre responses to complex pitch-producing stimuli. *Br. J. Audiol.* **21**, 311.

Evans, E.F. 1988 Cochlear nerve discharge patterns in response to complex stimuli: model predictions and neural data. *Br. J. Audiol.* **22**, 136.

A. J. M. HOUTSMA. Yes, we did collect some pitch matching data to complex tones of 20 harmonics and fundamentals around 50 Hz, i.e. upper edge frequencies between 920 and 1080 Hz. Partials were in sine phase, and complexes were at 10, 20, 40 and 60 dB above hearing threshold. Matches were performed by the authors, using sinusoidal comparison tones of constant intensity for all complex-tone levels.

The first preliminary finding was that the standard deviation of the matches was not greatly affected by complex-tone intensity, even close to threshold. This may appear surprising since Horst *et al.* (1985) showed that the FFT of period histograms of 8th-nerve-fibre responses to similar complex tones has 'rabbit ears' only at relatively high stimulus intensities.

A second and rather robust finding was that means of matches (see pitch shift shown in figure 3) increased systematically with stimulus intensity, from 35 Hz within the band at 10 dB SL to about 50 Hz outside the band at 60 dB SL, at a rate of about 1.5 Hz dB^{-1}. This finding appears entirely consistent with the prediction from physiological data you mentioned.

Reference

Horst, J.W., Javel, E. & Farley, G.R. 1985 Extraction and enhancement of spectral structure by the cochlea. *J. acoust. Soc. Am.* **78**, 1898–1901.

Perception of timbral analogies

STEPHEN McADAMS[1,2] AND JEAN-CHRISTOPHE CUNIBLE[1]

[1]*Laboratoire de Psychologie Expérimentale (CNRS URA 316), Université René Descartes, 28 rue Serpente, F-75006 Paris, France*
[2]*IRCAM, 31 rue Saint-Merri, F-75004 Paris, France*

SUMMARY

Recent studies have investigated the structure of perceptual relations among musical instrument timbres by multidimensional scaling (MDS) techniques. These studies have employed both acoustically produced tones and digitally synthesized imitations and hybrids of acoustic instrument tones. The analyses of dissimilarity ratings for all pairs of a set of tones are usually represented as geometrical structures in a two- or three-dimensional Euclidean space in which the shared 'perceptual' axes are shown to have a qualitative correspondence to acoustic properties such as spectral energy distribution, onset characteristics and degree of change in spectral distribution over the duration of the tone. The present study took as a point of departure a MDS analysis for complex, synthetic tones with the aim of testing whether musician and non-musician listeners used the relations defined by the perceptual space to perform an analogies task of the sort: timbre A is to timbre B as timbre C is to which of two possible timbres, D or D'? A parallelogram model was used to select the D timbres: if the relation between A and B is represented as a vector with both magnitude and direction components, then the appropriate D should form a vector with C having similar magnitude and direction in the timbre space. Aside from conceptual difficulties with the task for both non-musicians and composers, choices for both groups provide support for the parallelogram model indicating a capacity in listeners to perceive abstract relations among the timbres of complex sounds without specific training in such a task.

1. INTRODUCTION

One of the properties of the pitch dimension that endows it with its psychological capacity to serve as a vehicle for musical form is the fact that relations between pitches (i.e. intervals) can be perceived as musical qualities in their own right. Musical sequences can be built upon these qualities, and operations on musical material, such as transposition, that maintain them also maintain a strong degree of perceptual similarity between the original and transformed materials. If one were to try to extend the form-bearing possibilities of pitch into the realm of timbre, it would be necessary to determine the kinds of structuring of timbral relations that can be perceived by listeners and reasoned with by composers. Therefore, one of the important issues in research on musical timbre is the way in which listeners might potentially make use of relations among timbres in the perception of musical structure.

The trend toward using timbre in increasingly complex ways in music dates from orchestration practice in the last half of the 19th century (Boulez 1987). This trend has been extended considerably with the advent of analog and digital means of sound generation and processing. These same means provide the researcher with the possibility to generate with precise control sounds of considerable complexity and thus to open the way to the systematic study of timbre perception.

For the psychologist, several interesting questions arise concerning a listener's ability to perceive and remember timbral relations in tone sequences (Krumhansl 1989; McAdams 1989), as well as to build up hierarchical mental representations based these relations (Lerdahl 1987). Research in the past 20 years (cf. Plomp, 1970; Risset & Wessel 1982; Barrière 1990) has attempted to go beyond the loose negative definition of timbre given us by the field of psychoacoustics (i.e. timbre is what distinguishes two tones of identical pitch, loudness and perceived duration). To this end experimental paradigms that reveal the perceptual structure of timbral relations have been employed, and most notably those based on the multidimensional scaling of similarity (or dissimilarity) judgments.

In such a study, a number of tones differing in timbre (and equated for pitch, loudness, and perceived duration) are presented in all possible pairs to listeners who are asked to decide how dissimilar the tones of each pair are and to rate the dissimilarity on a scale of, say, 1 to 8. A multidimensional scaling algorithm is then applied to the matrix of judged dissimilarities. In many types of analyses, the algorithm tries to establish a monotonic relation between the dissimilarity ratings and Euclidean distances among the sounds arranged in a geometric structure in n dimensions, each sound being represented as a point. Sounds with similar timbres are thus near one another in the space and those with dissimilar timbres

Phil. Trans. R. Soc. Lond. B (1992) **336**, 383–389
Printed in Great Britain

are farther apart. The experimenter tries solutions with varying numbers of dimensions and selects the solution that is a compromise between having a small difference between distances and ratings (which decreases with increasing n) and not having more dimensions than can be readily interpreted in terms of their underlying perceptual and psychophysical relevance to the group of listeners tested. Different studies on timbre have generally settled on two (Plomp 1970; Wessel 1973, 1979; Ehresman & Wessel 1978; Rasch & Plomp 1982) or three dimensions (Grey 1977; Krumhansl 1989; Kendall & Carterette 1991). We will focus on the studies that adopted a three-dimensional solution for isolated timbres.

Grey (1977) used 16 digitally recorded, analysed resynthesized musical instrument tones performing a pitch at E^b_3 ($F_0 = 311$ Hz). Krumhansl (1989) used 21 synthetic tones developed by Wessel *et al.* (1987) on a Yamaha frequency modulation synthesizer: some of these tones were imitations of traditional Western orchestral instruments while others were hybrids (e.g. 'vibrone' is a hybrid of vibraphone and trombone, and 'guitarnet' is a hybrid of guitar and clarinet). Both Grey's and Krumhansl's spaces are qualitatively similar in the interpretation of their underlying dimensions, so we will confine our discussion to the latter since these tones were employed in our experiment.

A non-quantitative comparison of acoustic characteristics of the tones with their position along the various perceptual axes gave rise to the following interpretation (see figure 1). Dimension I seems related to the temporal envelope (rapidity of the attack and presence of inharmonic transients at the beginning of the tone) and might be called 'attack quality'. Sharp or biting attacks, such as that of the harpsichord, are found at one end of the dimension and softer, gentler attacks as with the clarinet are found at the other end. Dimension II seems related to a combined spectro-temporal property called 'spectral flux'. Instruments whose spectral envelope evolves relatively little over the duration of the tone (like the oboe) have low spectral flux compared to those whose spectrum changes a great deal (usually brightness increasing and decreasing with intensity as in the brass instruments). Dimension III seems related to the global spectral envelope and is called 'brightness'. Grey & Gordon (1978) have shown brightness to be highly correlated with the centre of gravity of the long-term spectrum represented in terms of specific loudness and critical band rate (Zwicker & Scharf 1965). Bright sounds (like the oboe) have a greater presence of energy in the higher harmonics than do less bright sounds (like the French horn). In most cases the hybrid instruments were situated between the two instruments from which they were derived.

An additional aspect of the Krumhansl (1989) analysis (based on a technique developed by Winsberg

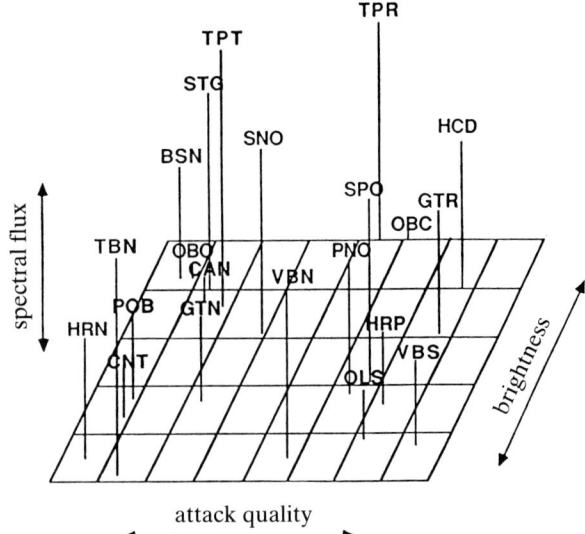

Figure 1. Timbre space derived from a three-dimensional scaling solution for dissimilarity judgments on 21 synthetic instrument tones. BSN = bassoon, CAN = cor anglais, CNT = clarinet, GTN = guitarnet (GTR/CNT), GTR = guitar, HCD = harpsichord, HRN = French horn, HRP = harp, OBC = obochord (OBO/HCD), OBO = oboe, OLS = oboleste (OBO/celeste), PNO = piano, POB = bowed piano, SNO = striano (STG/PNO), SPO = sampled piano, STG = string, TBN = trombone, TPR = trumpar (TPT/GTR), TPT = trumpet, VBN = vibrone (VBS/TBN), VBS = vibraphone. (Adapted from Krumhansl (1989).)

& Carroll, 1988)[†] revealed the existence of unique (although unspecified) perceptual features for certain instrument timbres. These features (called 'specificities' in the analysis technique) are not taken into account by the three common dimensions. Examples of specific features might include the odd-harmonic, hollow tone colour of the clarinet which is not subsumed under brightness, or the 'bump' at the return of the hopper on the end of a harpsichord tone. Eight of the 21 instruments had relatively high specificities (including the clarinet and harpsichord).

Once such a space has been quantified, one might ask whether the structure of the common dimensions is useful as a tool for predicting listeners' abilities to compare relations among the sounds. For example, can one use Euclidean spatial relations to define the properties of an interval formed by two timbres? This idea was initially developed by Ehresman & Wessel (1978) who applied Rumelhart & Abrahamson's (1973) parallelogram model of analogical reasoning in a semantic space to the timbre space composed of the tones used by Grey (1977). Rumelhart & Abrahamson took as a point of departure a three-dimensional space obtained by MDS techniques applied to dissimilarity judgments on animal names (Henley 1969). They were interested in whether the structure of the space would allow them to predict people's choices when presented with an analogy task of the form A is to B as C is to D (or A:B::C:D). In general, if the relation between two objects, A and B, is represented as a vector in the space, the model predicts that subjects will choose an object D which is the closest to the end point of a vector starting at C and having the

† In the MDS analysis with specificities, the algorithm tries to find a monotone relation between the dissimilarity ratings and estimated distances, d_{ij}, between the tones i and j, such that $d_{ij} = \{\Sigma (x_{ik} - x_{jk})^2 + s_i + s_j\}^{1/2}$, where x_{ik} is the coordinate on the kth dimension for tone i and s_i is the estimated specificity for tone i that is not accounted for by the common dimensions.

same magnitude and direction as **AB** (vectors are denoted in italic boldfaced type). They called this the ideal solution point, I. **AB** and **CI** thus form a parallelogram in the space. In their experiment, subjects were presented with analogies of the form A:B::C:{D_1, D_2, D_3, D_4}, where the D_i's varied according to their distance from I. The probability of choosing D_i as the best solution was found to be a monotonically decreasing function of the absolute distance of D_i from I, thus supporting the parallelogram model. Ehresman & Wessel proceeded in analogous fashion with musical instrument tones. The underlying assumption behind the definition of a timbre interval as a vector is that processes exist for the encoding and processing of relations between timbres that are isomorphic with those for representing and processing vector quantities. While the results were not as strongly supportive of the parallelogram model as in the Rumelhart & Abrahamson study, they were better predicted by this model than a number of other models. This early paper is encouraging as (i) it formalizes the notion of a timbre interval as being composed of both distance and degree of change along important perceptual dimensions, and (ii) it shows that this definition is correlated with listeners' judgments across intervals. The weakness of the study is that timbral vectors were computed from only a two-dimensional solution and that only relative vector magnitude was tested, ignoring the direction components. Our study systematically selected pairs of timbre vectors to be compared in an analogy task to test both magnitude and direction components.

2. METHOD

(a) Stimuli

Tones were derived from the set of 21 synthetic instruments described above. Each tone was realized playing an E^b_3 at mezzo forte (MIDI velocity 70) on a Yamaha TX802 FM Tone Generator. All sounds had been equalized for pitch and loudness by Krumhansl (1989). There were some significant differences in duration, however, certain plucked and struck sounds lasting longer than sounds imitating forced vibration instruments (winds and bowed strings). The nominal duration for each tone was 300 ms.

The magnitude and direction components of a vector between any pair of sounds in the three-dimensional perceptual space derived by Krumhansl for these tones can be computed as follows (e.g. for A and B).

Magnitude

Magnitude, d (corresponds to the estimated perceived dissimilarity):

$|\boldsymbol{AB}| = \{\Sigma(x_{Ak} - x_{Bk})^2\}^{1/2}$, where x_{Ak} is the coordinate on the kth dimension for timbre A, and $k = 1, 2, 3$.

Direction angles

Direction angles, α (degree of change on dimension I) and β (degree of change on dimension II; the angle

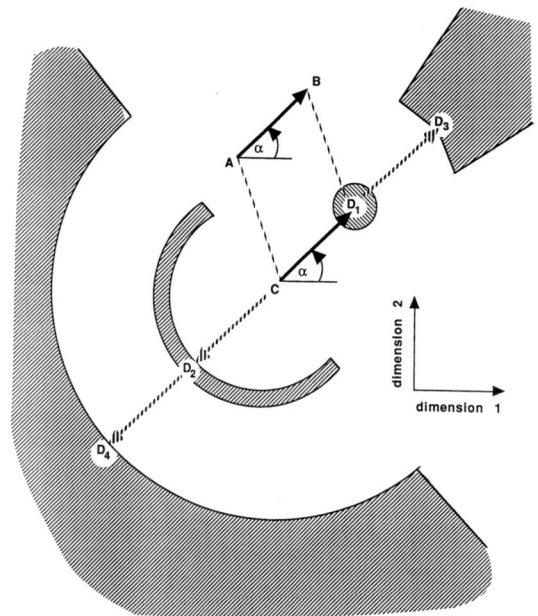

Figure 2. Two-dimensional representation of the different sequence types. The angle α is with respect to dimension 1. The angle β would be with respect to dimension 2 if the vectors were three-dimensional and coming out of the page. The hashed areas represent the constraint space for the end points of $\boldsymbol{CD}_i$ vectors and are labeled D_1, D_2, D_3 or D_4, accordingly. The ideal point I would be at the tip of the arrow-head for $\boldsymbol{CD}_i$. For the three-dimensional case, the area would be a sphere for D_1, a shell for D_2, part of a cone for D_3, and a solid with a spherical hollow for D_4.

γ for dimension III is complementary to the other two by the relation $\cos^2\alpha + \cos^2\beta + \cos^2\gamma \equiv 1$):

$$\alpha_{\boldsymbol{AB}} = \cos^{-1}((x_{A1} - x_{B1})/|\boldsymbol{AB}|),$$

$$\beta_{\boldsymbol{AB}} = \cos^{-1}((x_{A2} - x_{B2})/|\boldsymbol{AB}|).$$

Vectors can then be compared in terms of d, α and β. Accordingly, four classes of four-tone sequences were constructed to be of the form A:B::C:D_i. Constraints were established for the selection of four different kinds of D_i, such that the magnitude and direction components of **AB** and $\boldsymbol{CD}_i$ were either similar or quite different. These constraints are schematically illustrated (for the two-dimensional case only) in figure 2. They can be formalized as follows.

Sequence 1 – A:B::C:D_1 (right magnitude, right direction on **CD** with respect to **AB**); D_1 close to I with small error (ε) on d, α, and β:

$$|\boldsymbol{CD}_1| = |\boldsymbol{AB}| \pm \varepsilon_d,$$

$$\alpha_{CD1} = \alpha_{AB} \pm \varepsilon_\alpha,$$

$$\beta_{CD1} = \beta_{AB} \pm \varepsilon_\beta.$$

Sequence 2 – A:B::C:D_2 (right magnitude, wrong direction); small error on d, but at least one of α_{CD2} or β_{CD2} must differ by at least 90° from α_{AB} or β_{AB}, respectively:

$$|\boldsymbol{CD}_2| = |\boldsymbol{AB}| \pm \varepsilon_d,$$

$$|\alpha_{CD2} - \alpha_{AB}| \geqslant 90° \text{ and/or } |\beta_{CD2} - \beta_{AB}| \geqslant 90°.$$

[91]

Table I. *The possible sequence comparison types and the effects they were designed to test*

(Sequence labels are abbreviated: e.g. $D_2 = A:B::C:D_2$. Comparison D_2/D_3 was not included in the experiment because no pairs of sequences satisfying the appropriate constraints could be found in the chosen stimulus set.)

comparison type	vector component tested	origin of effect
D_1/D_2	direction	right magnitude in both cases right direction on D_1 wrong direction on D_2
D_1/D_3	magnitude	right direction in both cases right magnitude on D_1 wrong magnitude on D_3
D_1/D_4	magnitude and direction	right magnitude and direction on D_1 wrong magnitude and direction on D_4
D_2/D_3	magnitude vs. direction	right magnitude and wrong direction on D_2 wrong magnitude and right direction on D_3
D_2/D_4	magnitude under wrong direction	wrong direction in both cases right magnitude on D_2 wrong magnitude on D_4
D_3/D_4	direction under wrong magnitude	wrong magnitude in both cases right direction on D_3 wrong direction on D_4

Sequence 3 – $A:B::C:D_3$ (wrong magnitude, right direction) small error on α and β, but $|CD_3|$ must be larger than $|AB|$:

$$|CD_3| \geqslant 1.8|AB|,$$

$$\alpha_{CD3} = \alpha_{AB} \pm \varepsilon_\alpha,$$

$$\beta_{CD3} = \beta_{AB} \pm \varepsilon_\beta.$$

Sequence 4 – $A:B::C:D_4$ (wrong magnitude, wrong direction):

$$|CD_4| \geqslant 1.8|AB|,$$

$$|\alpha_{CD4} - \alpha_{AB}| \geqslant 90° \text{ and/or } |\beta_{CD2} - \beta_{AB}| \geqslant 90°.$$

In the above equations, the maximum allowed value of the error terms was fixed as follows: $|\varepsilon_d| \leqslant 0.35$, $|\varepsilon_\alpha| \leqslant 22.9°$, $|\varepsilon_\beta| \leqslant 22.9°$. These values were determined empirically to be as small as possible while giving a reasonable number of sequences for each type listed above.‡ The range of d for timbre pairs used in the experiment was 2.5–14.6 with a mean of 7.60. The range of angles was 14.2°–177.7° (mean = 95.7°) for α and 7.7°–164.6° (mean 104.8°) for β.

(b) Procedure

Ideally, we want to find appropriate D_1, D_2, D_3, and D_4 for any given set of A, B, and C tones and ask listeners to rank order them with respect to their relative success in fulfilling the analogy as was done in Ehresman & Wessel (1978). This would allow us to test directly for the relative importance of magnitude and direction components of the timbral vectors. With the given space however, this was impossible since sets of seven timbres (A, B, C, D_1, D_2, D_3, D_4) satisfying

‡ It should be noted that the accumulated error in ε_α and ε_β leads in some cases to an ε_γ as large as 86° which results in the D for that sequence being farther removed from I. The mean $|\varepsilon_\gamma|$ in D_1 and D_3 sequences was 36.3° with a standard deviation of 22.6°.

the constraints could not be found. We were obliged to settle on an experimental paradigm in which pairs of sequences were presented and subjects were to compare them and determine which best satisfied the analogy $A:B::C:D$. This reduced the stimulus search constraints to finding sets of five timbres (A, B, C, D, D'). The comparison types and the effect each was designed to test are listed in Table I. The following is an example of a D1/D4 comparison, where 'oboleste' is a hybrid of oboe and celeste:

D1: harp is to harpsichord as oboleste is to guitar, or
D4: harp is to harpsicord as oboleste is to clarinet.

At least five versions of each of the six possible pairs of sequence types were found with the exception of $A:B::C:D_2/A:B::C:D_3$ (subsequently referred to simply as D_2/D_3). This comparison was thus dropped from the experiment. Each version of a comparison was composed of different timbres while still satisfying the stimulus constraints for the two sequence types. The use of multiple versions allowed us to test the generality of the analogy task across different sets of timbres.

In each trial, listeners heard two sequences of four timbres with the following time structure, where the durations indicate silent intervals between the 300 ms tones: A–500 ms–B–900 ms–C–500 ms–D–1300 ms–A–500 ms–B–900 ms–C–500 ms–D'. After a pause of 2700 ms, the eight tone sequence was repeated once.

A complete block of 50 trials included the five sequence comparison types (D_1/D_2, D_1/D_3, D_1/D_4, D_2/D_4, D_3/D_4) each being presented in five versions with different timbres and with the order of presentation of the sequences counterbalanced.

Two groups of subjects were tested: 18 psychology students from René Descartes University without any formal musical training (non-musicians) and seven professional composers participating in a workshop on computer music at IRCAM. The non-musicians were tested individually over headphones in a single-walled soundproof chamber and entered their responses on the computer keyboard. The composers were tested in a group, listening to loudspeakers in a sound treated studio and entered their responses on a numbered answer sheet. The non-musicians completed two blocks of trials whereas the composers completed a single block. The sounds were presented at a comfortable listening level.

Subjects were given an instruction sheet that explained the analogy task using a semantic and a visual example. The correct solutions to each example were explained. Six practice trials were given using a randomly selected set of experimental trials. No feedback was given on either the practice or the experimental trials. After completing the practice trials, any further questions the subject(s) had were answered before proceeding to the first block of trials.

(c) *Hypotheses*

The following sets of hypotheses were tested in the experiment. Each hypothesis refers to a separate aspect of the data. They are thus not mutually exclusive.

1. Subjects will prefer D_1 over D_2, D_3, and D_4 as a solution to the analogy, as it is the best fit to the parallelogram model. A corollary to this hypothesis would predict that the preference of D_1 over D_4 be stronger than that over D_2 or D_3 as D_4 is the farthest removed in all respects from the ideal point.
2. D_2 will be preferred over D_4: listeners prefer the right magnitude even though the direction is wrong in both **CD** intervals.
3. D_3 will be preferred over D_4: listeners prefer the right direction even though the magnitude is wrong in both **CD** intervals.
4. There will be no differences among the different versions of each comparison type since the analogy judgment is based on a perception of abstract relations among the timbres of the stimulus tones.
5. The effects of hypotheses 1–3 will be stronger for composers than for non-musicians as the activity of reasoning with sound and making timbre judgments in composition will have allowed the former group to develop more consistent judgment strategies.

An additional point of interest concerns the missing D_2/D_3 condition. In the absence of this condition, a comparison between D_1/D_2 and D_1/D_3 preferences will indicate something of the relative effect of distance and direction. We have no *a priori* hypothesis about this result based on the parallelogram model.

3. RESULTS AND DISCUSSION

The data consisted of percent choices of one of the paired sequences over the other for each version of each comparison type collected across order of presentation. An effect of block of trials was only found for the D_1/D_2 comparison in the nonmusicians' data: the percent choice of D_1 was greater in the second block

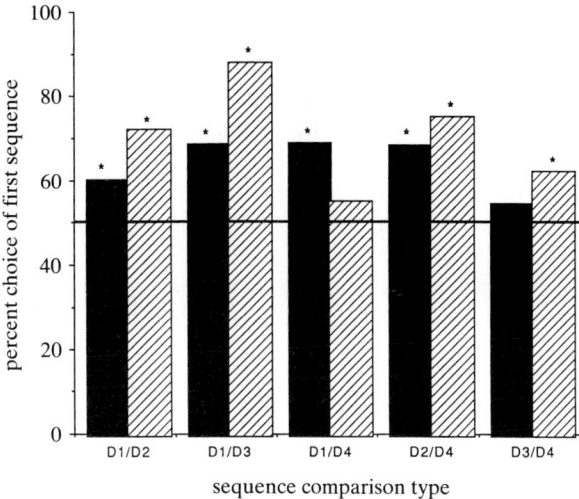

Figure 3. Global means (across versions, presentation orders and listeners) for five sequence comparison types. The comparison type is labelled on the horizontal axis. The two groups of subjects (18 non-musicians and seven composers) are shown with solid and hashed bars, respectively. The horizontal line is positioned at 50% (chance choice). The asterisks over certain bars indicate that the mean is significantly different from chance.

(two-tailed, $t_{(17)} = 3.01$, $p < 0.01$). In the subsequent analyses, the data are grouped across blocks for the nonmusicians.

The means for the experimental conditions are highly correlated between subject groups ($r = 0.65$, $p < 0.01$). Although composers tend to express stronger preferences (one-tailed, $t_{(24)} = 1.66$, $p = 0.055$), the patterns of both data sets are qualitatively similar. Thus hypothesis 5 is at most only weakly supported by the data.

(a) *Global effect of comparison type*

The means for each comparison type obtained for each subject group are shown in figure 3. To test for differences from chance choice (50%), one-group t-tests were performed on means for each comparison type across versions for each of the subject groups. The Bonferoni-adjusted criterion was 0.005 (ten tests). All means except for the D_3/D_4 comparison were significantly different from chance for nonmusicians, and all except for the D_1/D_4 comparison were different from chance for composers.

Hypothesis 1 which predicted that D_1 would be preferred over all other sequences is confirmed in all cases for non-musicians and in all cases except D_1/D_4 for composers. This latter result is quite surprising, because according to the parallelogram model, D_4 should be the farthest from the ideal point and D_1 the closest. Examination of the means for the five versions of D_1/D_4 for composers shows that three hover around chance, one is significantly higher than chance (preference for D_1), and one is quite lower than 50% (preference for D_4) although this latter mean just misses being significantly different from 50%. In general, however, the results suggest that the parallelogram model captures a significant portion of subjects' judgment strategies since the timbre closest to the ideal point is preferred over other more distant timbres. The corollary to hypothesis 1 is not confirmed, i.e. preferences for D_1 over D_4 are not higher than those of D_1 over D_2 or D_3. This will require further reflection since the parallelogram model of Rumelhart & Abrahamson (1973) predicts monotonically decreasing preference with increasing distance from the ideal point.

That relative distance between timbre pairs can be evaluated by listeners even though the directions are dissimilar is suggested by the fact that the mean preference for D_2 over D_4 is reliably above chance for both subject groups. Hypothesis 2 is thus confirmed indicating that the distance component of the timbral change is perceptually important in perceiving timbral relations.

Hypothesis 3 (D_3 is preferred over D_4) is confirmed for composers but not for non-musicians. This result suggests that the former group can evaluate relative direction of timbral change even though the distances of the two D timbres from C are quite different from that between A and B. Examination of the five versions of D_3/D_4 for the non-musicians reveals that two had means reliably above 50% (preference for D_3) and one was significantly below 50% (preference for D_4).

In the absence of a D_2/D_3 condition, a comparison of D_1/D_2 means with those for D_1/D_3 suggests that distance change across timbre pairs (D_1/D_3) is more easily noticed than direction change (D_1/D_2), because D_1 is preferred more over D_3 than over D_2. This difference is not statistically significant, however, indicating that neither magnitude nor direction predominates over the other component.

Overall the results are encouraging, witnessing an ability to make judgments on timbral relations. However, some of these global effects need to be qualified by a closer look at the different versions grouped under each comparison type.

(b) Effects of individual versions of each comparison type

To test for effects of individual versions within comparison type, one-way analyses of variance with repeated measures on version were performed. The results are shown in table 2. For both subject groups, four out of five comparison types have significant overall differences between versions. This indicates that not every version of each comparison had the same perceptual result and was thus not judged in a similar way. In particular, one notes a great dispersion of means for certain comparisons (D_3/D_4 for non-musicians and D_1/D_4 for composers). This dispersion on either side of 50% results in the global mean being not different from random choice. Globally, we must reject hypothesis 4 which predicted equal performance for all versions of a comparison type.

(c) Effect of the relative distance of D_is from the ideal point

According to the Rumelhart & Abrahamson model, the choice of one sequence over another should be a monotonically increasing function of the distance between the ideal point and D. Therefore, for each comparison type, these distances were calculated and the mean percent choices for each comparison type were regressed onto the difference between these distances. This analysis indicates the degree to which judgments may have been based purely on the relative distance of D from I in each sequence. The regression was performed independently for nonmusician and composer groups. For non-musicians the regression yielded a significant fit between mean data and distances ($R = 0.48$; $F_{(1,23)} = 6.80$, $p < 0.05$). Although the fit is not bad, the regression only accounts for 23% of the variance in the data indicating that other factors are entering into the judgments that are unaccounted for by a simple distance-from-ideal-point model. For composers, the fit between mean data and distances is not significant ($R = 0.04$; $F_{(1,23)} = 0.04$). In spite of the strong correlation between the means for non-musicians and composers, there appears to be no relation between relative distance from the ideal point and the sequence preferred as best completing the analogy for the composers.

Another possibility is that listeners made judgments based on the relative degree of change along the individual perceptual dimensions. Accordingly, we performed a multiple regression of the differences in change along each dimension between **AB** and **CD** or **CD'** vectors onto mean percent choice for each group of listeners. For nonmusicians, the fit was not significant ($R = 0.46$; $F_{(3,21)} = 1.93$) whereas for composers the fit was significant ($R = 0.57$; $F_{(3,21)} = 3.33$, $p < 0.05$). The partial Fs for the multiple regression show that differences in change along dimensions I and II (attack, spectral flux) are largely responsible for this fit. Taken together, these two regression analyses may indicate differences in listening and judgment strategies between the two groups.

4. CONCLUSIONS

A number of experimental conditions were designed within the framework of a Euclidean distance model of timbre space (Krumhansl, 1989) in order to test

Table 2. *One-way analyses of variance with repeated measures on version for each comparison type and subject group*

(Sequence labels are abbreviated: e.g. $D_2 =$ A:B::C:D_2. For non-musicians $n = 18$, and for composers $n = 7$.)

comparison type	non-musicians		composers	
	$F_{(4,68)}$	p	$F_{(4,24)}$	p
D_1/D_2	3.58	< 0.01	8.26	< 0.005
D_1/D_3	2.36	> 0.05	3.12	> 0.10
D_1/D_4	4.49	< 0.005	5.14	< 0.005
D_2/D_4	9.20	< 0.001	7.10	< 0.001
D_3/D_4	9.00	< 0.001	3.88	< 0.05

listeners' abilities to perceive timbral relations and to judge their similarity in terms of magnitude and direction of timbre change. These results support and extend those of Ehresman & Wessel (1978). They are also coherent with work by Kendall & Carterette (1991) who have shown that a vector model of timbre can account for perceived similarities among simultaneously sounding wind instrument dyads.

A vector model of timbre intervals was fairly successful at predicting the choice of one type of sequence over another, where the sequences varied in the degree to which the magnitude and direction components of the timbral vectors match across pairs of timbres. In general, timbres close to the ideal point predicted by the vector model were preferred as best fulfilling an analogy of the form A:B::C:D than were timbres that were at some distance from that point (conditions D1/D2, D1/D3, D1/D4). We have also shown that in some cases the model even predicts preference when both Ds in a sequence comparison are quite far removed from I, indicating an ability to appreciate the appropriate vector magnitude under conditions of wrong direction (D2/D4) and of appropriate direction under conditions of wrong magnitude (D3/D4), though the latter condition is quite weak. What the model does not do is make predictions about the relative contributions of magnitude and direction of the comparison timbre vector. This is a subject for future research.

The strong effect of the timbre set chosen to realize each comparison type suggests a relative lack of generalizability of timbral interval perception across different timbres. This result may be due to a number of factors that were not controlled in this study: (i) there may be a relative instability of judgment strategies, since most of the listeners have never encountered a listening situation in which focusing on, or comprehending, abstract timbral relations was appropriate; (ii) there may be effects of the relative magnitude of a given vector and the distance between to-be-compared vectors: very large vectors may be difficult to compare with precision and small vectors that are very far apart in the space may also be difficult to compare; (iii) there may be effects of the degree of change along different common dimensions: the perceptual weights of change along individual dimensions may not be equivalent in this kind of listening task; and (iv) there may be effects of specific features of individual timbres that are not taken into account by the common dimensions of the timbre space, but which influence the perceived distances between timbres and thus the timbre intervals that are to be compared.

Portions of this study were realized in partial fulfillment of the requirements for J.-C. Cunibile's Master's thesis at the Laboratoire de Psychologie Expérimentale, Université René Descartes (Cunibile 1991). This research was supported in part by a grant from the French Ministry of Culture.

REFERENCES

Barrière, J.-B. (ed.) 1990 *Le timbre, métaphore pour la composition*. Paris: Christian Bourgois/IRCAM.

Boulez, P. 1987 Timbre and composition—timbre and language. *Contemp. Music Rev.* **2**, 161–172.

Cunibile, J.-C. 1991 Perception des analogies de timbre. Master's thesis, Laboratoire de Psychologie Expérimentale, Université René Descartes, Paris.

Ehresman, D. & Wessel, D.L. 1978 Perception of timbral analogies. *Rapp. IRCAM*, no. 13. Paris: IRCAM.

Grey, J.M. 1977 Multidimensional perceptual scaling of musical timbre. *J. acoust. Soc. Am.* **61**, 1270–1277.

Grey, J.M. & Gordon, J.W. 1978 Perceptual effects of spectral modifications on musical timbres. *J. acoust. Soc. Am.* **63**, 1493–1500.

Henley, N.M. 1969 A psychological study of the semantics of animal terms. *J. verb. Learn. verb. Behav.* **8**, 176–184.

Kendall, R.A. & Carterette, E.C. 1991 Perceptual scaling of simultaneous wind instrument timbres. *Music Percept.* **8**, 369–404.

Krumhansl, C.L. 1989 Why is musical timbre so hard to understand? In *Structure and perception of electroacoustic sound and music* (ed. S. Nielzen & Olsson), pp. 43–53. Amsterdam: Elsevier (Excerpta Medica 846).

Lerdahl, F. 1987 Timbral hierarchies. *Contemp. Music Rev.* **2**, 135–160.

McAdams, S. 1989 Psychological constraints on form-bearing dimensions in music. *Contemp. Music Rev.* **4**, 181–198.

Plomp, R. 1970 Timbre as a multidimensional attribute of complex tones. In *Frequency analysis and periodicity detection in hearing* (ed. R. Plomp & G. F. Smoorenburg), pp. 397–414. Leiden: Sijthoff.

Rasch, R. & Plomp, R. 1982 The perception of musical tones. In *The psychology of music* (ed. D. Deutsch), pp. 1–24. New York: Academic Press.

Risset, J.-C. & Wessel, D.L. 1982 Exploration of timbre by analysis and synthesis. In *The psychology of music* (ed. D. Deutsch), pp. 25–58. New York: Academic Press.

Rumelhart, D.E. & Abrahamson, A.A. 1973 A model for analogical reasoning. *Cogn. Psychol.* **5**, 1–28.

Wessel, D.L. 1973 Psychoacoustics and music. *Bull. comput. Arts Soc.* **30**, 1–2.

Wessel, D.L. 1979 Timbre space as a musical control structure. *Comput. Music J.* **3**(2), 45–52.

Wessel, D.L., Bristow, D. & Settel, Z. 1987 Control of phrasing and articulation in synthesis. *Proceedings of the 1987 International Computer Music Conference*, pp. 108–116. San Francisco: Computer Music Association.

Winsberg, S. & Carroll, J.D. 1988 A quasi-nonmetric method for multidimensional scaling via an extended Euclidean model. *Psychometrika* **53**, 217–229.

Zwicker, E. & Scharf, B. 1965 A model of loudness summation. *Psychol. Rev.* **72**, 3–26.

Some new pitch paradoxes and their implications

DIANA DEUTSCH

Department of Psychology, University of California, San Diego, La Jolla, California 92093, U.S.A.

SUMMARY

This paper explores two new paradoxical sound patterns. The tones used to produce these patterns consist of six octave-related harmonics, whose amplitudes are scaled by a bell-shaped spectral envelope; these tones are clearly defined in terms of pitch class (C, C#, D, and so on) but are poorly defined in terms of height. One pattern consists of two tones that are separated by a half-octave. It is heard as ascending when played in one key, yet as descending when played in a different key. Further, when the pattern is played in any one key it is heard as ascending by some listeners but as descending by others (the tritone paradox). Another pattern that consists of simultaneous pairs of tones displays related properties (the semitone paradox). It is shown that the way the tritone paradox is perceived correlates with the speech characteristics of the listener, including his or her linguistic dialect. The findings suggest that the same, culturally acquired representation of pitch classes influences both speech production and also perception of this musical pattern.

1. INTRODUCTION: PITCH CLASS AND PITCH HEIGHT

Pitch, as defined by the American National Standards Institute (1973) is 'that attribute of auditory sensation in terms of which sounds may be ordered on a scale extending from high to low'. Indeed, most experimental research on the subject has assumed such a definition. However, it has long been recognized by musicians that pitch is not a unidimensional attribute; rather it varies along two dimensions: the monotonic dimension of pitch height and the circular dimension of pitch class (Westergaard 1975). Tones that are related by octaves; i.e. whose fundamental frequencies stand in the ratio of 2:1, are in some sense perceptually equivalent. Indeed, the system of notation for the traditional musical scale assumes such equivalence. The core of this scale consists of twelve tones, which are formed by the division of the octave into equal logarithmic steps, called semitones. Each tone is assigned a name (C, C#, D, and so on), and the entire scale is produced by repeatedly traversing the circle of note names across successive octaves.

Assuming, then, that pitch varies along two dimensions, the question arises as to whether these dimensions are orthogonal, or whether they interact in some fashion. Shepard (1964) performed an experiment which addressed this issue. He generated a series of tones, each of which consisted of ten harmonic components which were separated by octaves. The amplitudes of these components were scaled by a fixed, bell-shaped spectral envelope, and the pitch classes of the tones were varied by shifting the components up and down in log frequency, keeping the position and shape of the envelope constant. Subjects were presented with ordered pairs of such

tones, and they reported in each case whether they heard an ascending or a descending pattern. When the tones within a pair were separated by one or two steps along the pitch class circle, judgments were almost entirely determined by proximity. For example, the pattern D-D# was always heard as ascending, and the pattern C#-C was always heard as descending. As the distance between the tones along the pitch class circle increased, the tendency for judgments to be determined by proximity lessened, and when the tones were separated by exactly a half-octave, ascending and descending judgments occurred equally often.

Shepard concluded from these findings that the dimensions of pitch class and pitch height are indeed orthogonal. However, this conclusion can be questioned on two grounds. First, since judgments in this study were largely determined by proximity, any influence of pitch class on perceived height could have been overwhelmed by this factor (Deutsch 1982). Second, the judgments were presented as averaged both over subjects and also over pitch classes, so that any influence of pitch class on perceived height would have been lost in the averaging process.

The foregoing considerations led the present author to embark on a series of experiments to determine whether an influence of pitch class on perceived height would be manifest for patterns in which proximity could not be used as a cue. The patterns used were composed of octave-related complexes similar to those employed by Shepard, and the data were analysed for each subject separately as a function of the pitch classes of the tones comprising each pattern. Using this procedure, the perceived heights of the tones were found to vary in an orderly fashion as a function of their positions along the pitch class circle, so that tones in one region of the circle were heard as higher and

Phil. Trans. R. Soc. Lond. B (1992) **336**, 391–397
Printed in Great Britain

391

© 1992 The Royal Society

[97]

tones in the opposite region were heard as lower. Furthermore, the direction of this relationship varied substantially across subjects, so that any given pattern was heard by different listeners in strikingly different ways.

2. EQUIPMENT AND TONE COMPLEXES

Tones were generated on a VAX 11/780 computer, interfaced with a DSC-200 Audio Data Conversion System (16 bit, 48K sampling rate). They were recorded and played back on a Sony PCM-F1 Digital Audio Processor, and the output was passed through a Crown amplifier and delivered to subjects binaurally through headphones (Grason-Stadler TDH-49) at a level of approximately 72 dB SPL.

Each tone consisted of six sinusoids which were separated by octaves, the amplitudes of which were determined by a fixed, bell-shaped spectral envelope. The following is the general form of the equation describing the envelope:

$$A(f) = 0.5 - 0.5 \cos\left[\frac{2\pi}{\gamma}\log_\beta\left(\frac{f}{f_{min}}\right)\right] f_{min} \leqslant f \leqslant \beta^\gamma f_{min}$$
$$A(f) = 0 \text{ elsewhere}$$

where $A(f)$ is the relative amplitude of a given sinusoid at frequency f Hz, β is the frequency ratio formed by adjacent sinusoids (thus for octave spacing, $\beta = 2$), γ is the number of β cycles spanned, and f_{min} is the minimum frequency for which the amplitude is

non-zero. The maximum frequency for which the amplitude is non-zero is thus $\gamma\beta$ cycles above f_{min}. The values $\beta = 2$ and $\gamma = 6$ were used throughout, so that the spectral envelope always spanned exactly six octaves, from f_{min} to $64 f_{min}$.

3. THE TRITONE PARADOX

One-pattern that was explored consisted of an ordered pair of tones which were separated by a half-octave, or tritone (Deutsch 1986). Because the tones were diametrically opposed along the pitch class circle, proximity could not here be used as a cue in determining their relative heights. On each trial, subjects were presented with one such tone pair, and they judged whether it formed an ascending or a descending pattern; from these judgments it was inferred which tones were heard as higher and which as lower.

All tones were 500 ms in duration, and there were no gaps between tones within a pair. To control for possible effects based on the relative amplitudes of the harmonic components, and also to examine the effects of variations in overall height, the tone pairs were generated under envelopes that were placed at six different positions along the spectrum. The envelope peaks were spaced at half-octave intervals, and were at C_6 (1047 Hz), $F\#_5$ (740 Hz), C_5 (523 Hz), $F\#_4$ (370 Hz), C_4 (262 Hz), and $F\#_3$ (185 Hz). All twelve pitch-class pairings (C-F#, C#-G, D-G#, . . ., B-F) were presented equally often under each of the six positions of the spectral envelope.

Figure 1 shows the results from two subjects, in each

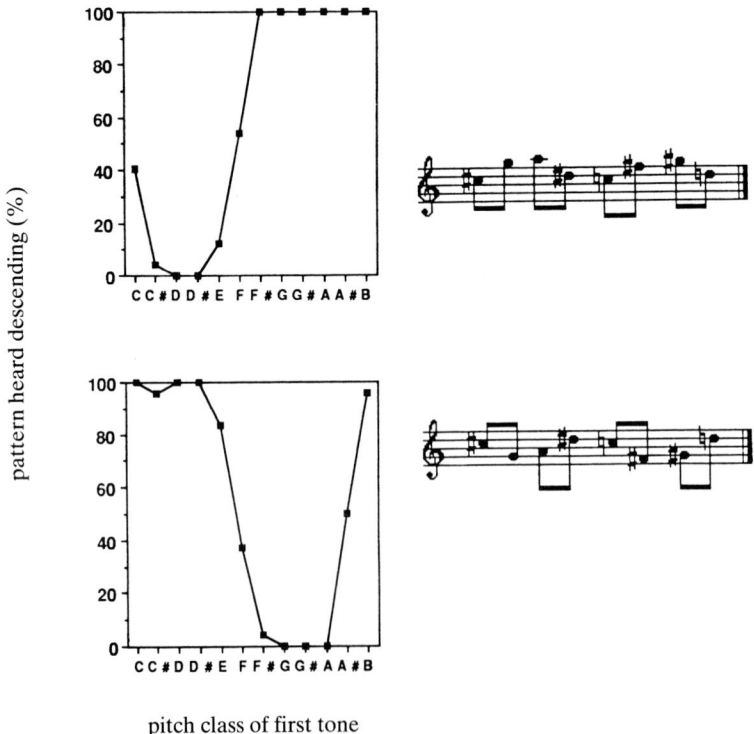

Figure 1. The tritone paradox. Graphs on left show percentages of judgments that a tone pair formed a descending pattern, plotted as a function of the pitch class of the first tone of the pair. The results from two subjects are here displayed, avaraged over all six positions of the spectral envelope, and over two experimental sessions. Musical notations on right show how these two subjects perceived the identical series of tone pairs C#-G, A-D#, C-F#, G#-D. (Data from Deutsch (1986).)

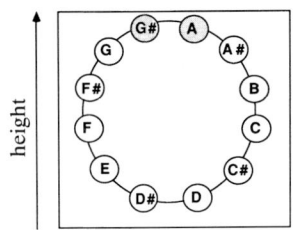 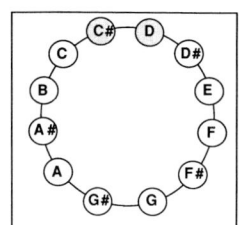

Figure 2. Perceptual orientations of the pitch class circle with respect to height, derived from the judgments of the two subjects shown in figure 1. Filled circles indicate the pitch classes that defined the highest position along the circle; these are termed peak pitch classes.

case averaged over two experimental sessions, and over all six positions of the spectral envelope. It can be seen that for both subjects, judgments depended in an orderly fashion on the positions of the tones along the pitch class circle. However, the direction of this dependence differed strikingly between the subjects: for the most part, when the first subject heard an ascending pattern the second subject heard a descending one, and vice versa. In consequence, extended passages composed of such tone pairs were heard by these two subjects as forming entirely different melodic contours. An example is given in the righthand part of figure 1.

Figure 2 depicts the two perceptual orientations of the pitch class circle with respect to height, which are derived from the judgments of two subjects shown in figure 1. For the first subject, the pitch classes that defined the highest position along the circle (hereafter referred to as peak pitch classes) were C# and D instead.

Deutsch (1987) investigated the effects of the spectral envelope position in detail. The data from four subjects were examined, and twelve envelope positions were employed: these varied in one-quarter octave steps, and so over a three octave range. It was found that in all cases judgments depended systematically on the positions of the tones along the pitch class circle, and also that the direction of this dependence varied from one subject to another. Although there were some interactions between pitch class and the relative amplitudes of the sinusoidal components, and also between pitch class and overall height, these interactions were not necessarily present. Further, even when interactions were found, their effects in absolute terms were generally quite small.

In a large-scale study (Deutsch *et al.* 1987), a group of subjects was selected on the only criteria that they were undergraduates at the University of California, San Diego, that they had normal hearing, and that they could judge reliably whether pairs of single sinusoids that were separated by a half-octave formed ascending or descending patterns. The subjects made judgments of the tritone paradox, and the results were analysed as follows. It was first determined for the scores of each subject separately whether the pitch class circle could be bisected in such a way that none of the scores in the upper half of the circle was lower than any of the scores in the lower half. This criterion was fulfilled by 22 of the 29 subjects. Next, the

proportion of random permutations of the scores that could be so characterized was determined by computer simulation. Averaged across subjects, this proportion was found to be 0.027 per subject. The probability of obtaining the combined result by chance was thus shown to be vanishingly small, and it can be concluded that the influence of pitch class on perceived height occurs to a highly significant extent in this general population.

In this study, the peak pitch classes were tabulated for each subject, and the distribution of peak pitch classes within the subject population was determined. An orderly bell-shaped distribution was obtained, so that for the most part pitch classes B, C, C#, D, and D# were heard as higher, and pitch classes F, F#, G, G#, and A were heard as lower. This finding will be further considered below.

4. THE SEMITONE PARADOX

We next consider what happens when more than one tone is presented at a time. To examine this, a pattern was created which consisted of two simultaneously presented pairs of tones; one pair formed a pattern that ascended by a semitone whereas the other formed a pattern that descended by a semitone (Deutsch 1988). The tone pairs were diametrically opposed along the pitch class circle, so that, again, proximity could not be used as a cue in making judgments of their relative heights. All tones were 500 ms in duration, and there were no gaps between tones within a pair. In preliminary work it was found that this pattern was generally perceived as two stepwise lines that moved in contrary motion. However, for any given instantiation of the pattern, some listeners heard the higher line as ascending and the lower line as descending, whereas other listeners heard the higher line as descending and the lower line as ascending. In the formal study, patterns were generated under envelopes which were placed at 12 different positions along the spectrum, which were spaced at one-quarter octave intervals. Under each position of the spectral envelope, each of the 12 pitch classes served equally often as the first tone of an ascending pair, and also as the first tone of a descending pair.

Four subjects were employed in this study, and it was found that, again, judgments reflected orderly relationships between the perceived heights of the tones and their positions along the pitch class circle. Also, again, the direction of the relationship between pitch class and perceived height varied considerably across subjects. This is exemplified in the judgements of two subjects shown in figure 3. It can be seen that as the pattern was transposed, the ascending and the descending lines appeared to interchange positions. However, for the most part, when the first subject heard the higher line as ascending the second subject heard it as descending, and vice versa. Thus extended passages composed of such patterns were perceived by these two subjects in quite different ways. This is exemplified in the passage shown on the righthand part of figure 3. Concerning the effects of envelope position, it was found that, as in the study of Deutsch (1987), there were some interactions between pitch class and the relative amplitudes of the sinusoidal

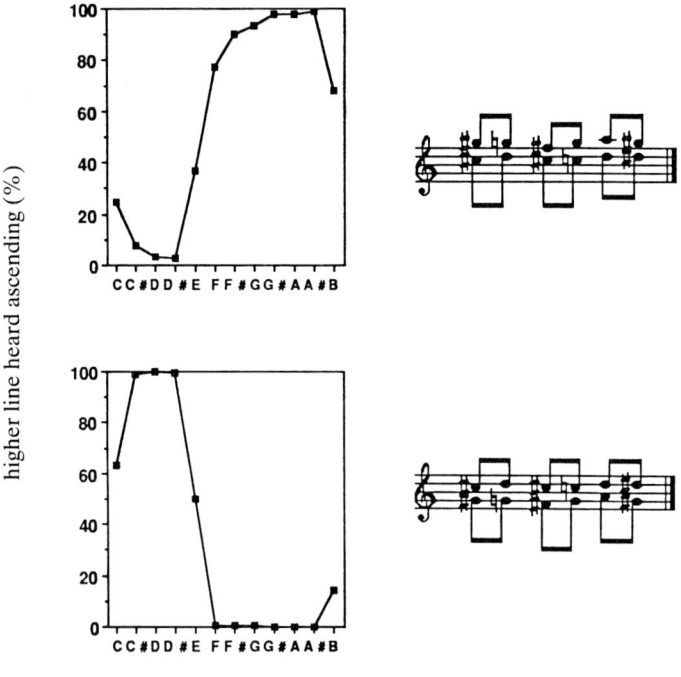

tones in ascending line

Figure 3. The semitone paradox. Graphs on left display the percentages of judgments that the higher line formed an ascending pattern, plotted as a function of the pitch classes of the tones in the ascending line. The results from two subjects are here displayed, averaged over twelve positions of the spectral envelope, and over nine experimental sessions. Musical notations on right show how these two subjects perceived the identical series of patterns G#-G/ C#-D, F#-G/C#-C, A-G#/D-D#. (Data from Deutsch (1988).)

components, and there were also some interactions between pitch class and overall height. However, these interactions were not necessarily present, and even when they were, their effects in absolute terms were generally quite small.

5. BASIS OF THE TRITONE PARADOX

As described earlier, Deutsch *et al.* (1987) found that the tritone paradox occurred in the large majority of subjects in a sizeable population, showing that this phenomenon is not confined to a few selected individuals. Within this population, no correlate with musical training was found, either in terms of the probability of obtaining the phenomenon, or its size, or its direction. A number of other studies have also ruled out explanations in terms of low-level properties of the hearing mechanism. For many subjects, the profiles relating pitch class to perceived height were largely unchanged when the position of the spectral envelope was shifted over a three octave range (as, for example, in the study of Deutsch (1987)). In a recent experiment, the tritone paradox was produced when the odd-numbered components of each tone complex were presented to one ear and the even-numbered components were presented to the other ear. However, the effect was not produced when either the odd-numbered components or the even-numbered components were presented alone to both ears. Since the phenomenon can result from integration of information presented separately to the two ears, then it follows that it must be central in origin (Deutsch 1992a)†.

A number of informal observations led the author to hypothesize that perception of the tritone paradox might be related to the processing of speech sounds. Specifically, it was hypothesized that through long-term exposure to such sounds, the listener acquires a representation of his or her own speaking voice, and that included in this representation is a delimitation of the octave band in which the largest proportion of pitch values occurs. It was further hypothesized that the pitch classes delimiting this octave band for speech are taken as defining the highest position along the pitch class circle (i.e. the position bounded by the peak pitch classes, as sown in figure 2), and that this in turn determines the perceived orientation of the pitch class circle with respect to height‡.

---

† Space does not permit a discussion of the reasons why the tritone paradox did not occur with tone complexes whose components were spaced at two-octave intervals. One possibility is that the subjects were here using spectral proximity in making their judgments. However, the explanation of this negative result is irrelevant to the point made here, that the tritone paradox can be produced by central combination of information presented to each ear separately.

‡ It may be noted that the spectral compositions of the tones used here to produce the tritone paradox differed substantially from those of speech sounds. However, the author recently performed a study using pairs of single tones, each of which comprised a full harmonic series, but in which the relative amplitudes of the odd and even harmonics were such as to produce ambiguities of perceived height. At least for some subjects, the tritone paradox was still obtained, and the judgments made were very similar to those made by the same subjects with the use of octave-related complexes (Deutsch 1992b). The conditions giving rise to the tritone paradox may therefore be rather general, provided that the presented sounds are such as to produce ambiguities of perceived height.

Table 1. *Pitch classes delimiting the octave band for speech, together with those defining highest position along the pitch class circle, tabulated by subject (from Deutsch et al. 1990)*

subject	limit of octave band for speech	highest position along pitch class circle	distance in semitones
A.H.	D#-E	D-D#	1
D.M.	D-D#	D#-E	1
D.D.	F-F#	G-G#	2
T.T.	D#-E	D-D#	1
M.D.	E-F	G#-A	4
M.C.	A#-B	C-C#	2
M.M.	D#-E	D-D#	1
E.S.	C-C#	C-C#	0
W.B.	D#-E	D#-E	0

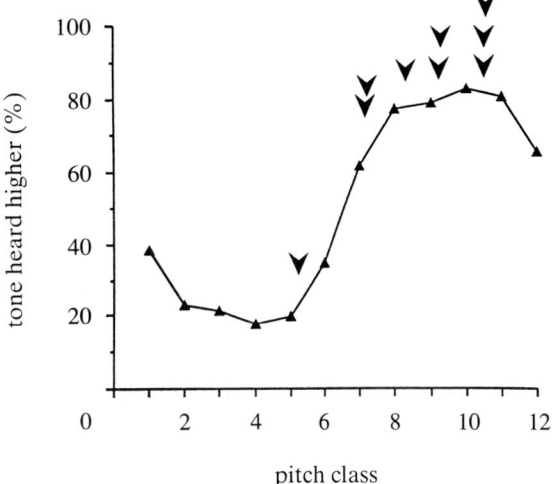

Figure 4. Percentages of trials in which a tone was heard as the higher of a pair in making judgments of the tritone paradox, with the orientation of the pitch class circle normalized and averaged across subjects. Arrows indicate the limit of each subject's octave band for speech, in relation to the highest position along the pitch class circle, as determined by judgments of the tritone paradox. (From Deutsch *et al.* (1990).)

Deutsch *et al.* (1990) performed an experiment to test this hypothesis. Nine subjects (four male and five female) were selected who showed clear and consistent relationships between pitch class and perceived height in making judgments of the tritone paradox. A 15 min recording of spontaneous speech was taken from each subject, and the speech samples were recorded into computer memory. F_0 estimates were then obtained from these speech samples at 4 ms intervals. These estimates were allocated to semitone bins, and histograms were derived of the percentage occurrence of F_0 estimates in each bin. From each histogram, the octave band containing the largest number of F_0 estimates was determined. Finally, for each subject, comparison was made between the pitch classes delimiting this octave band for speech, and those defining the highest position along the pitch class circle, as determined by judgments of the tritone paradox. As shown in table I, for eight of the nine subjects, the values were separated by no more than two semitones, so that a significant correspondence between these values was established ($p = 0.04$, two-tailed, on a binomial test). Figure 4 displays a further representation of these findings, in which the limit of each subject's octave band for speech is plotted on a curve showing the judgments of the tritone paradox, normalized and averaged across all subjects§.

The findings of Deutsch *et al.* (1990) are in accordance with the above hypothesis, which we may then consider in two versions. The first version makes no assumption that the listener's pitch range for speech is itself determined by a learned template. The second version assumes that such a template is acquired through exposure to speech produced by others, and that it is employed both to evaluate perceived speech

§ Because the correlate here was between pitch classes rather than frequency values, the statistical significance of this correlate did not depend on the hypothesis that the listener focuses on the upper limit of the octave band for speech in generating a pitch class template. However, it appears intuitively reasonable that the listener would indeed focus on the upper end rather than the lower one, as it is here that speech sounds are most salient. It should further be noted that the choice of an octave band follows logically from the findings on the tritone paradox, since this paradox is a pitch class phenomenon. The peak of the F_0 distribution in the speech histogram is not an appropriate measure here, as it lies somewhere in the middle of the distribution.

and also to constrain the listener's own speech production. On this second hypothesis, the characteristics of such a template would be expected to be similar for people who speak in the same language or dialect, but to vary for people who speak in different languages or dialects.

The second hypothesis was tested in an experiment by Deutsch (1991). Two groups of subjects made judgements of the tritone paradox. The first group consisted of 24 subjects who had grown up in California, and the second consised of 12 subjects who had grown up in the South of England. None of the subjects in the Californian group had a parent who had grown up in England, and none of the subjects in the English group had a parent who had grown up in California.

For each subject, the percentage of judgments that a tone pair formed a descending pattern was plotted as a function of the pitch class of the first tone of the pair. The distribution of peak pitch classes was then determined for the Californian and English groups separately. These distributions are shown in figure 5, and it can be seen that they are strikingly different: for the Californian group, B, C, C#, D, and D# occurred most often as peak pitch classes, but for the English group, F#, G, and G# occurred most often instead.

To evaluate the statistical significance of the difference between the two groups, the hypothesis was tested that the distribution produced by the Californian group would be similar to that obtained earlier in the study of Deutsch *et al.* (1987) on undergraduates at the University of California, but that the distribution produced by the English group would be a different one. Comparison was therefore made between the number of subjects in each group for whom the peak position along the pitch class circle lay in the half of the circle containing the larger number of peak positions in the study of Deutsch *et al.* (1987). It was found that 21 of the 24 Californian subjects fell

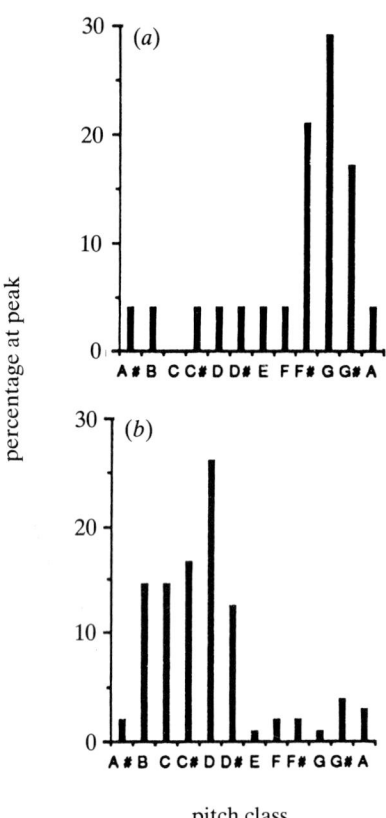

Figure 5. Distribution of peak pitch classes (see figure 2) as determined by judgments of the tritone paradox, in two groups of subjects. The first group (a) had grown up in the South of England, and the second (b) had grown up in California. (From Deutsch (1991).)

into this category; however, only three of the 12 English subjects did so. The difference between the two groups on this measure was therefore highly significant ($p < 0.001$ on a Fisher exact probability test). These findings are therefore in accordance with the hypothesis of a culturally acquired representation of the pitch class circle which is specific to the language or dialect of the listener.

What can be the evolutionary value of such a template? As one line of reasoning, it could be advantageous to determine the emotional state of a speaker through the pitch of his or her voice. A template such as hypothesized here could provide a framework, common to a particular dialect, within which the pitch of a speaker's voice may be evaluated, and this would in turn provide evidence concerning his or her emotional state. It may be observed that a template that is based on pitch class rather than pitch height has the useful feature that it can be invoked for both male and female speakers, even though their speech involves different pitch ranges. Such a template could also be invoked in communicating syntactic aspects of speech‖.

‖ It is interesting to note that, in retrospect, subjects who had shown large differences in the orientation of the pitch class circle with respect to height in making judgments of the tritone paradox had frequently come from different linguistic backgrounds. For example, the subject whose data are shown in the upper portion of figure 1 had grown up in London, and the subject whose data are shown in the lower portion of this figure had grown up in California.

The findings of Deutsch *et al.* (1990) and Deutsch (1991) taken together lead to the prediction that, for a large sample of individuals in a given linguistic group, the distribution of fundamental frequencies in spontaneous speech for males and females should stand in octave relation. It should be emphasized, however, that based on the data from Californians obtained by Deutsch (1991) the spread should be quite large (almost a half-octave). Further, the correlate between the peak pitch classes derived from judgments of the tritone paradox and the pitch classes delimiting the octave band for speech found by Deutsch *et al.* (1990) was obtained on subjects who were selected for showing clear and consistent functions in making judgments of the tritone paradox. On might expect, therefore, that in an unselected sample of subjects the correlate would be less clear.

6. FURTHER IMPLICATIONS

One further implication of these findings concerns theories of pitch perception. Two factors have previously been shown to contribute to the perceived pitch of a complex tone. First, when presented with a series of harmonics the listener perceives a pitch that corresponds to the frequency of the fundamental. Second, the listener assigns weightings to the harmonics on the basis of their relative amplitudes, and perceives a pitch in accordance with these weightings. The first factor is held to predominate for tones with fundamentals below roughly 900 Hz, and the second factor for tones with fundamentals above this value (Goldstein 1973; Scharf & Houtsma 1986; Terhardt *et al.* 1982). The present findings show that, at least under certain circumstances, a third factor operates also: the perceived height of a tone can be influenced by its position along the pitch class circle.

Another implication of the findings concerns musical transposition. It is generally taken as self-evident that when a passage is transposed from one key to another, the perceived relationships between the pitches are unchanged. The patterns explored here are striking cases in which this principle is violated: For the tritone paradox, transposition results in a perceived inversion; for the semitone paradox, transposition results in a perceived interchange of voices.

A further implication concerns the phenomenon of absolute pitch, which is characterized as the ability to name a note that occurs in isolation. In general, this faculty is considered very rare. However, the present paradoxes demonstrate that the large majority of individuals possess a form of absolute pitch, in that we perceive tones as higher or as lower depending solely on their pitch classes, or note names.

A final implication concerns relationships between speech and music. It has been acknowledged since antiquity that these two modes of communication have characteristics in common; however the bases for such commonalities have so far been undetermined. The findings reported here provide, to the author's knowledge, the first indication that one form of communication can exert a direct influence over the other, and this opens the door to the possible uncovering of other such influences.

REFERENCES

American National Standards Institute 1973 *American national psychoacoustical terminology*. S3.20. New York: Anonymous.

Deutsch, D. 1982 Grouping mechanisms in music. In *The psychology of music* (ed. D. Deutsch), (pp. 99–134). New York: Academic Press.

Deutsch, D. 1986 A musical paradox. *Mus. Percept.* **3**, 275–280.

Deutsch, D. 1987 The tritone paradox: effects of spectral variables. *Percept. Psychophys.* **41**, 563–575.

Deutsch, D. 1988 The semitone paradox. *Mus. Percept.* **6**, 115–132.

Deutsch, D. 1991 The tritone paradox: an influence of language on music perception. *Mus. Percept.* **8**, 335–347.

Deutsch, D. 1992*a* The tritone paradox and central integration. (In preparation.)

Deutsch, D. 1992*b* The tritone paradox produced by full harmonic series. (In preparation.)

Deutsch, D., Kuyper, W.L. & Fisher, Y. 1987 The tritone paradox: Its presence and form of distribution in a general population. *Mus. Percept.* **5**, 79–92.

Deutsch, D., North, T. & Ray, L. 1990 The tritone paradox: Correlate with the listener's vocal range for speech. *Mus. Percept.* **7**, 371–384.

Goldstein, J.L. 1973 An optimum processor theory for the central formation of the pitch of complex tones. *J. acoust. Soc. Am.* **54**, 1496–1516.

Scharf, B. & Houtsma, A.J.M. 1986 Audition II: Loudness, pitch, localization, aural distortion, pathology. In *Handbook of perception and human performance*, vol II, 15 (ed. K. Boff, L. Kaufmann & J. Thomas), pp. 1–60. New York: Wiley.

Shepard, R.N. 1964 Circularity in judgments of relative pitch. *J. acoust. Soc. Am.* **36**, 2345–2353.

Terhardt, E., Stoll, G. & Seewann, M. 1982 Pitch of complex signals according to virtual pitch theory: tests, examples, and predictions. *J. acoust. Soc. Am.* **71**, 671–678.

Westergaard, P. 1975 *An introduction to tonal theory*. New York: Norton.

Coding of envelope modulation in the auditory nerve and anteroventral cochlear nucleus

XIAOQIN WANG† AND MURRAY B. SACHS

Center for Hearing Sciences and Departments of Biomedical Engineering and Otolaryngology-Head and Neck Surgery, Johns Hopkins University, Baltimore, Maryland 21205, U.S.A.

SUMMARY

We have investigated responses of the auditory nerve fibres (ANFs) and anteroventral cochlear nucleus (AVCN) units to narrowband 'single-formant' stimuli (SFSS). We found that low and medium spontaneous rate (SR) ANFs maintain greater amplitude modulation (AM) in their responses at high sound levels than do high SR units when sound level is considered in dB SPL. However, this partitioning of high and low SR units disappears if sound level is considered in dB relative to unit threshold. Stimuli with carrier frequencies away from unit best frequency (BF) were found to generate higher AM in responses at high sound levels than that observed even in most low and medium SR units for stimuli with carrier frequencies near BF.

AVCN units were shown to have increased modulation depth in their responses when compared with high SR ANFs with similar BFs and to have increased or comparable modulation depth when compared with low SR ANFs. At sound levels where AM almost completely disappears in high SR ANFs, most AVCN units we studied still show significant AM in their responses. Using a dendritic model, we investigated possible mechanisms of enhanced AM in AVCN units, including the convergence of inputs from different SR groups of ANFs and a postsynaptic threshold mechanism in the soma.

1. INTRODUCTION

Natural sounds like speech can be thought of as combinations of narrowband stimuli (Flanagan 1972). The waveform of a narrowband signal is characterized by two major features: a carrier frequency and an envelope. The temporal patterns of auditory nerve fibre (ANF) spike trains can be phase-locked both to carrier frequency, and to the envelope of narrowband stimuli like amplitude-modulated sounds (Moller 1976; Rose *et al.* 1967). The phase locking to stimulus envelope is an important phenomenon, because it reveals properties of the cochlear filter that can not be studied with pure tone stimuli, but also has implications for possible mechanisms underlying the extraction of the pitch of complex stimuli from ANF firing patterns (De Boer 1976; Schouten 1940).

A detailed description of ANF responses to narrowband stimuli is also relevant to the analysis of signal transformations in the cochlear nucleus. For example, low spontaneous rate (SR) ANFs have higher thresholds and wider dynamic ranges for rate responses to tones than do high SR ANFs (Liberman 1978; Sachs *et al.* 1989). It has been suggested that spectral features of complex stimuli are represented in the average discharge rate of low SR ANFs at high sound levels, where the rates of low threshold, high SR ANFs are saturated

(Delgutte 1982; Winslow *et al.* 1987). At low sound levels below the thresholds of low SR ANFs, spectral features are represented in the discharge rate of the high SR ANFs (Sachs & Young 1979). Blackburn & Sachs (1990) present evidence that chopper units in the anteroventral cochlear nucleus (AVCN) may 'listen selectively' to high SR ANF inputs at low sound levels and to low SR ANF inputs at high sound levels. However, details of the processing are not known. Evidence suggests that both low and high SR ANFs contact stellate cells (Liberman 1992; Ryugo *et al.* 1992), the source of chopper responses (Rhode *et al.* 1983), but we know little about how stellate cells integrate inputs from different ANF SR groups. One way to investigate such problems is to establish a physiological 'marker' that can distinguish low and high SR fibres at all sound levels and then study this marker in chopper units. We will show that the envelope fluctuations of temporal discharge patterns of responses to narrowband sounds can be used as such a marker.

2. METHODS

Single-unit recordings were made from nembutal anaesthetized cats. Surgical procedures were the same as those reported previously (Blackburn & Sachs 1989; Sokolowski *et al.* 1989) and approved by the Johns Hopkins Animal Care and Use Committee. Figure 1a illustrates generation of the narrowband

† Present address: Department of Physiology, University of California, Box 0732, San Francisco, California 94143-0732, U.S.A.

Phil. Trans. R. Soc. Lond. B (1992) **336**, 399–402
Printed in Great Britain

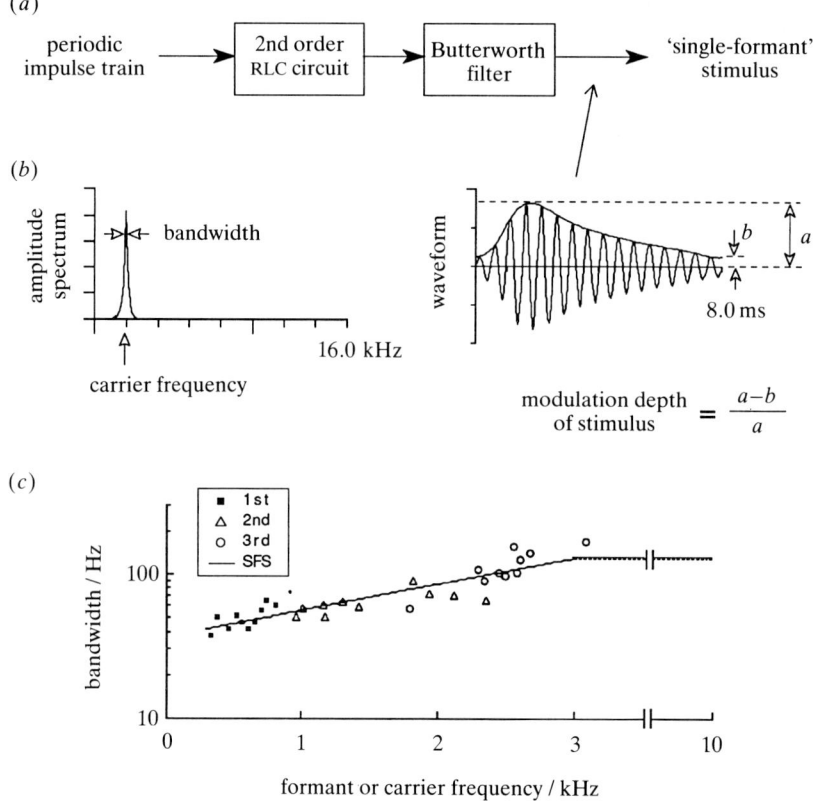

Figure 1. (*a*) Illustration of the process to generate 'single-formant' stimuli (SFSs). (*b*) An example of an SFS whose carrier frequency is 2.0 kHz. Both the stimulus waveform and its amplitude spectrum are plotted. Superimposed on the waveform is the computed envelope. (*c*) Plot of 3 dB bandwidth of formants of vowel sounds measured by Dunn (1961) (filled squares, first formant; open triangles, second formant; open circles, 3rd formant) and those of the SFSs (solid line) versus formant or carrier frequency.

'single-formant' stimuli (SFSs) used in this study. An example of an SFS with carrier frequency equal to 2.0 kHz is shown in figure 1*b*. The stimuli were digitally synthesized to approximate the response of an RLC circuit to a periodic impulse train. For each unit studied, the carrier frequency of the stimulus was usually set as close as possible (always within 1%) to the unit's best frequency (BF); in some cases responses to off-BF carriers were studied. The bandwidth of the stimulus was set equal to the average bandwidth of vowel formants at the stimulus carrier frequency. The formant bandwidth versus frequency function was taken from Dunn (1961) for frequencies less than 3.0 kHz, and set to a constant at higher frequencies (figure 1*c*). The fundamental frequency of the stimulus was the subharmonic of the carrier frequency closest to 125 Hz. For each unit, the appropriate SFS was presented at a number of sound levels. Period histograms were constructed from the responses and the envelopes extracted from the histograms. Details about computation of period histograms and their envelopes can be found in Wang (1991). The amplitude modulation (AM) in the SFSs and neural responses is measured by modulation depth, which is defined in figure 1*b*.

We separated ANFs into low SR (SR < 18.0 spikes per second) and high SR (SR ⩾ 18.0 spikes per second) groups. The low SR group in this presentation includes both low and medium SR ANFs as defined in many

other studies because our data do not show significant difference between low and medium SR ANFs in coding SFSs. AVCN units were classified into six types using a classification scheme similar to that used by Blackburn & Sachs (1989).

3. RESPONSES OF AUDITORY NERVE FIBRES

For stimuli centered at unit BF, responses of low-BF ANFs to SFSs are phase-locked to carrier frequency and to stimulus envelope; responses of high-BF units are phase-locked only to stimulus envelope. Period histograms from two low-BF ANFs are shown in figure 2 (*a–d*), with computed envelopes superimposed. The amount of modulation in the envelope of the period histogram is a function of sound level, increasing as sound level exceeds BF threshold and decreasing dramatically at high sound levels (Wang 1991). Period histograms at 40 dB SPL (figure 2*a*, *c*) show greater modulation than do those at 70 dB SPL (figure 2*b*, *d*). Low SR ANFs maintain higher envelope modulation at higher sound levels (dB SPL) than do high SR ANFs as can be seen by comparing period histograms of the high SR ANF to those of the low SR ANF in figure 2. This difference is a direct consequence of the higher thresholds of low SR ANFs (Wang 1991) and suggests that information about envelope modulation is carried by low SR ANFs at high sound levels.

low sound level
(40 dB SPL)

high sound level
(70 dB SPL)

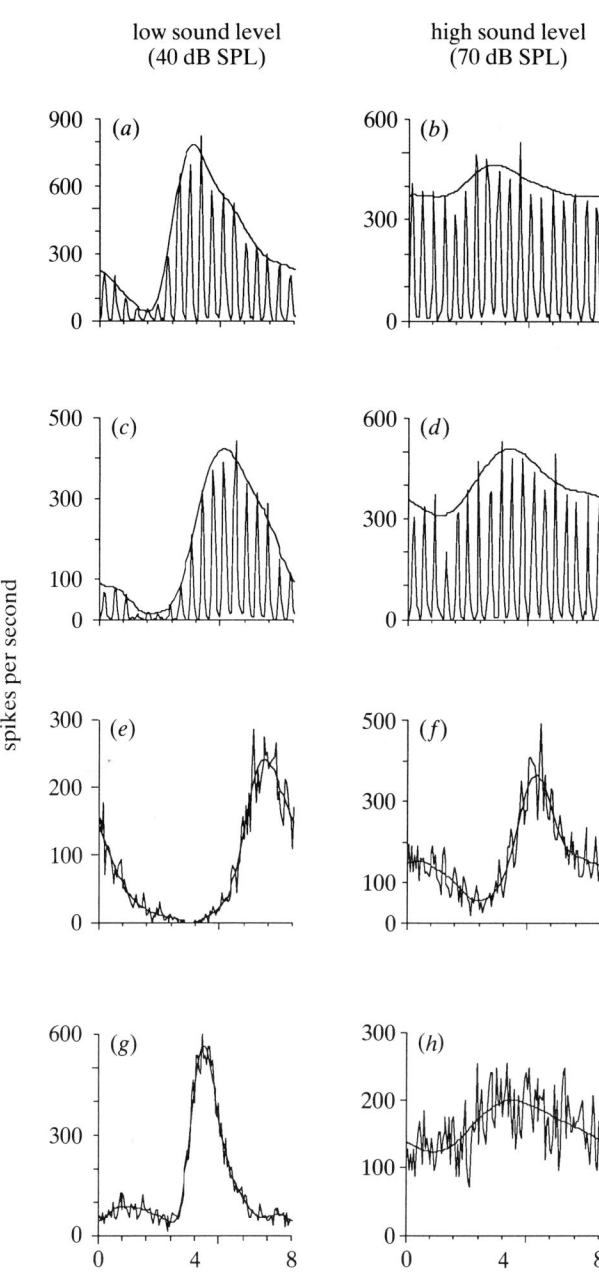

spikes per second

time / ms

time / ms

Figure 2. Period histograms with superimposed envelopes from (*a*, *b*) a high SR ANF (unit 1.05, 8/10/1990, BF 2.215 kHz, SR 51.0 spikes per second), (*c*, *d*) a low SR ANF (unit 2.01, 17/9/1990, BF 2.215 kHz, SR 2.2 spikes per second), (*e*, *f*) an AVCN chopper unit (unit 7.03, 7/8/1990, BF 2.215 kHz), and (*g*, *h*) simulated chopper responses. All auditory units have the same BF (2.215 kHz). Period histograms plotted in the left column are at a low sound level (40 dB SPL), whereas those in the right column are at a high sound level (70 dB SPL). For simulated chopper period histograms in (*g*) and (*h*), 10 ANF inputs at the soma were used. The inputs for the simulated period histograms shown in (*g*) were from the high SR ANF at the low sound level (shown in (*a*)); inputs in (*h*) were from the low SR ANF at the high sound level (shown in (*d*)).

Stimuli with carrier frequencies centered away from unit BF were found to generate large AM in responses at high sound levels, larger than that observed even in most of low SR ANFs for stimuli with carrier frequencies

near BF (data not shown). This result reflects both the higher thresholds for off-BF stimuli as well as a much more gradual decrease in modulation with increasing sound level above threshold. Therefore, off-BF inputs to cells in AVCN provide another potential source of envelope modulation at high sound levels.

4. RESPONSES OF AVCN UNITS

We studied AVCN units using the same stimulus used in the study of ANFs described above. Figure 2*e*, *f* show period histograms with envelopes, from a chopper unit whose BF is equal to that of the ANFs in figure 2*a–d*. Note that the amount of modulation at 70 dB SPL in the chopper unit (figure 2*f*) is much higher than that in either of the ANFs at the high sound level. The period histograms from this chopper unit show no phase locking to the carrier, as is typical of chopper units at this BF (2.215 kHz; Blackburn & Sachs 1989). Period histograms from primarylike units (recorded from AVCN bushy cells) show phase locking to the carrier comparable to that of ANFs. On the other hand primarylike units also show greater phase locking to the envelope than do ANFs at high sound levels (Wang 1991). In fact, all types of AVCN units show higher or comparable modulation depth at moderate to high sound levels than do ANFs (Wang 1991). At sound levels where AM almost completely disappears in high SR ANFs, most of AVCN units studied still show significant AM in their responses.

5. MECHANISMS OF MODULATION ENHANCEMENT

The enhanced modulation depth observed in AVCN units can result from several mechanisms. Based on our analysis of ANF and AVCN responses to SFSS, we suggest that these mechanisms include: (i) convergence of both low and high SR ANFs onto an AVCN cell; (ii) convergence of inputs from ANFs with BFs different from that of the AVCN cell; and (iii) the threshold effect due to temporal summation of subthreshold excitatory postsynaptic potentials (EPSPs) and inhibitory inputs at soma. These mechanisms were examined in a dendritic model of chopper units originally developed by Banks & Sachs (1991). Spike times of ANF responses to SFS were used as inputs to the model whose output was then compared with real chopper responses to SFSS. Two period histograms generated by the model are shown in figure 2*g*, *h*. The histogram in figure 2*g* shows the model response with inputs generated from the high SR ANF at low sound level (40 dB SPL; figure 2*a*). The histogram in figure 2*h* was generated with inputs from the low SR ANF at the high sound level (70 dB SPL; figure 2*d*). In both cases, ten inputs converge on the soma. We found that the peak to trough amplitude of the envelope of model period histograms is proportional to the amount of envelope modulation in the input. Thus if the high SR ANF at 70 dB SPL (figure 2*b*) is used as the input to the model, model output shows very little envelope modulation (Wang 1991).

Responses of the model with only low SR inputs

[107]

(figure 2*h*) show little enhancement of modulation relative to that in the inputs. When inputs consist of both low and high SR ANFs at high sound levels, the modulation depth of the model output is lower than that obtained without high SR ANFs. To compensate for the decrease in modulation depth due to high SR ANFs, which have almost no modulation in discharge patterns at high sound levels, and to achieve higher modulation depth in model output than that of low SR ANFs, a postsynaptic threshold mechanism is needed. This mechanism must work in such a way that a chopper cell gives output spikes only when input spike rate exceeds a threshold value. We implemented such a threshold mechanism in the model by modifying the steady-state activation and inactivation curves of the model spike generator on the basis of physiological observations (Oertel 1983; Smith & Rhode 1989). The model period histogram shown in figure 2*h* was obtained using the original model as described in Banks & Sachs (1991). A simulation using the modified model with the same ANF inputs does produce increased modulation depth (Wang 1991). Simulated inhibitory input to the soma, which effectively raises discharge threshold, also increases envelope modulation in the period histogram (data not shown).

REFERENCES

Banks, M.I. & Sachs, M.B. 1991 Regularity analysis in a compartmental model of chopper units in the anteroventral cochlear nucleus. *J. Neurophysiol.* **65**(3), 606–629.

Blackburn, C.C. & Sachs, M.B. 1989 Classification of unit types in the anteroventral cochlear nucleus: post-stimulus time histograms and regularity analysis. *J. Neurophysiol.* **62**, 1303–1329.

Blackburn, C.C. & Sachs, M.B. 1990 The representation of the steady-state vowel sound /e/ in the discharge patterns of cat anteroventral cochlear nucleus neurons. *J. Neurophysiol.* **63**(5), 1191–1212.

De Boer, E. 1976 On the residue and auditory pitch perception. *Handbook of sensory physiology*, vol. V (*Auditory system, part 3: Clinical and special topics*). New York: Springler-Verlag.

Delgutte, B. 1982 Some correlates of phonetic distinctions at the level of the auditory nerve. In *The representation of speech in the peripheral auditory system* (ed.), pp. 131–149. Amsterdam: Elsevier Biomedical Press.

Dunn, H.K. 1961 Methods of measuring vowel formant bandwidths. *J. acoust. Soc. Am.* **33**, 1737–1746.

Flanagan, J. 1972 *Speech analysis, synthesis and perception.* New York: Springer-Verlag.

Liberman, M.C. 1978 Auditory-nerve responses from cats raised in a low-noise chamber. *J. acoust. Soc. Am.* **63**, 442–455.

Liberman, M.C. 1992 Central projections of auditory-nerve fibers of differing spontaneous rate, I: anteroventral cochlear nucleus. Salamanca, Spain: NATO Advanced Research Workshop.

Moller, A.A. 1976 Dynamic properties of primary auditory fibers compared with cells in the cochlear nucleus. *Acta physiol. scand.* **98**, 157–167.

Oertel, D. 1983 Synaptic responses and electrical properties of cells in brain slices of the mouse anteroventral cochlear nucleus. *J. Neurosci.* **3**, 2043–2053.

Rhode, W.S., Oertel, D. & Smith, P.H. 1983 Physiological response properties of cells labeled intracellularly with horseradish peroxidase in cat ventral cochlear nucleus. *J. comp. Neurol.* **213**, 448–463.

Rose, J.E., Brugge, J.F., Anderson, D.J. & Hind, J.E. 1967 Phase-locked response to low frequency tones in single auditory nerve fibers of the squirrel monkey. *J. Neurophysiol.* **30**, 269–293.

Ryugo, D.K., Wright, D.D. & Pongstaporn, T. 1992 Ultrastructural analysis of synaptic endings of auditory nerve fibers in cats: correlations with spontaneous discharge rate. Salamanca, Spain: NATO Advanced Research Workshop.

Sachs, M.B., Winslow, R.L. & Sokolowski, B.A.H. 1989 A computational model for rate-level functions from cat auditory-nerve fibers. *Hear. Res.* **41**, 61–70.

Sachs, M.B. & Young, E.D. 1979 Encoding of steady-state vowels in the auditory nerve: Representation in terms of discharge rate. *J. acoust. Soc. Am.* **66**, 470–479.

Schouten, J.F. 1940 The residue and the mechanisms of hearing. *Proc. K. ned. Akad. Wet.* **43**, 991–999.

Smith, P.H. & Rhode, W.S. 1989 Structural and functional properties distinguish two types of multipolar cells in the ventral cochlear nucleus. *J. comp. Neurol.* **282**, 595–616.

Sokolowski, B.A.H., Sachs, M.B., Goldstein, J.L. 1989 Auditory nerve rate-level functions for two-tone stimuli: possible relation to basilar membrane nonlinearity. *Hear. Res.* **41**, 15–124.

Wang, X. 1991 Neural encoding of single-formant stimuli in the auditory-nerve and anteroventral cochlear nucleus of the cat. Ph.D thesis, The Johns Hopkins University.

Winslow, R.L., Barta, P.E. & Sachs, M.B. 1987 Rate coding in the auditory nerve. In *Auditory processing of complex sounds*, pp. 212–224. Hillsdale, New Jersey: Lawrence Erlbaum Assoc.

Discussion

E. F. EVANS (*Department of Communication and Neuroscience, University of Keele, U.K.*). In respect of the hypothetical threshold device, could inhibition be an adequate mechanism for this?

X. WANG. Our simulations with a compartment model showed that somatic inhibition can be an adequate mechanism for the proposed threshold device.

Modelling the sensitivity of cells in the anteroventral cochlear nucleus to spatiotemporal discharge patterns

LAUREL H. CARNEY

Department of Biomedical Engineering, Boston University, 44 Cummington Street, Boston, Massachusetts 02215, U.S.A.

SUMMARY

This study investigates a potential mechanism for the processing of acoustic information that is encoded in the spatiotemporal discharge patterns of auditory nerve (AN) fibres. Recent physiological evidence has demonstrated that some low-frequency cells in the anteroventral cochlear nucleus (AVCN) are sensitive to manipulations of the phase spectrum of complex sounds (Carney 1990*b*). These manipulations result in systematic changes in the spatiotemporal discharge patterns across groups of low-frequency AN fibres having different characteristic frequencies (CFs). One interpretation of these results is that these neurons in the AVCN receive convergent inputs from AN fibres with different CFs, and that the cells perform a coincidence detection or cross-correlation upon their inputs. This report presents a model that was developed to test this interpretation.

1. INTRODUCTION

It has been known for several decades that the temporal discharge patterns of low-frequency auditory nerve (AN) fibres convey information about the spectrum of an acoustic stimulus (see, for example, Kiang *et al.* (1965); Rose *et al.* (1967)). For responses to pure tones, the interpretation of temporal response patterns is straightforward: the dominant periodicity of a histogram of the response represents the frequency of the tonal stimulus. However, for complex sounds, the interpretation of temporal information is not so simple: periodicities within the responses of single fibres do not uniquely encode a peak frequency in a stimulus spectrum (Carney & Yin 1988). This complication arises because the temporal responses of AN fibres to complex sounds are influenced by a combination of the spectral properties of the stimulus and the filtering properties of the basilar membrane. However, because these filtering properties change systematically as a function of characteristic frequency (CF), by combining information encoded across a population of fibres of different CFs the central nervous system could accurately interpret information encoded in the temporal discharges. Thus, when considering the temporal encoding of complex sounds, we must consider not single fibre discharge patterns but rather spatiotemporal discharge patterns; that is, temporal patterns of activity across populations of fibres varying in CF.

The first opportunity for the processing of spatiotemporal discharge patterns arises in the anteroventral cochlear nuclear (AVCN), where AN fibres converge upon single cells (see Osen 1970). Convergence of

individual fibres onto postsynaptic neurons provides a potential neural substrate for coincidence detection; that is, a postsynaptic cell will in general be most likely to discharge when its inputs discharge simultaneously. The only exception occurs when activity of a single presynaptic input is capable of eliciting a postsynaptic discharge.

The processing of incoming information performed by a cell depends upon both the nature of its inputs and the cell's biophysical properties. For a cell in the AVCN: (i) the size and location of each input terminal will influence whether it can drive the cell alone or only when summed with other coincident inputs; (ii) the distribution of CFs and the number of inputs will play a role in determining the sensitivity of the cell to different input patterns; and (iii) the electrical properties of the cell's membrane, which influence its ability to summate inputs spatially and temporally, will determine the sensitivity of the cell to changes in the relative discharge times of its inputs.

Cells in the AVCN are known to display diversity in many of these aspects. At one extreme are the large spherical bushy cells, in the most rostral pole of the AVCN, that receive a single AN input in the form of a large somatic terminal (for examples, see Osen (1970); Lorente de No (1981)). In this case, each and every input discharge is thought to evoke a postsynaptic discharge (Pfeiffer 1966). At the other extreme are the stellate cells that receive many small AN terminals on the dendrites and the cell body (see Osen 1970). This study focused on the globular bushy cells, which receive convergent input from a few to several AN inputs including the relatively large modified end-

Phil. Trans. R. Soc. Lond. B (1992) **336**, 403–406
Printed in Great Britain

403

bulbs of Held (Osen 1970; Liberman 1991). Many of these cells have low spontaneous discharge rates and some have been shown to have subthreshold excitatory postsynaptic potentials *in vivo* (Smith & Rhode 1987). This result implies that, at least in some cases, individual input fibres are not capable of driving the cell to discharge, and so coincident inputs may be necessary to drive the cell. Furthermore, globular bushy cells have nonlinear membrane properties that reduce their ability to summate their inputs temporally (Wu & Oertel 1984), and thus their performance as coincidence detectors should be enhanced.

We do not know the relative CFs of the inputs that converge upon single cells in the AVCN, yet this feature will determine the functional significance of coincidence detection. In a previous study, it was shown that some cells with globular bushy response types were sensitive to manipulations of spatiotemporal patterns across frequency (Carney 1990*b*). An interpretation of that result is (i) that these cells receive input from fibres of different CFs, and (ii) that they perform a crosscorrelation upon their inputs. To understand better the implications of a cross-correlational mechanism for processing of complex sounds, we developed a model for globular bushy discharge patterns. We can use this model in future studies to explore the effects of the number of inputs, the distribution of the CFs of the inputs, and the coincidence requirements of the postsynaptic cell, as well as to identify ways to estimate these parameters in future physiological experiments.

2. METHODS

Inputs to the AVCN cell model were produced by a recently developed model for the responses of low-frequency AN fibres (Carney 1990*a*, 1992). The AN model takes arbitrary complex stimulus waveforms, and consists of four main stages: (i) a time-varying band-pass filter, representing the mechanical tuning of the basilar membrane with a compressive nonlinearity; (ii) a memoryless saturating nonlinearity followed by low-pass filtering, representing inner hair cell transduction and membrane properties; (iii) synaptic adaptation; and (iv) spike generation with refractoriness.

The AVCN cell was modelled by a simple shot-noise threshold model similar to those of Colburn *et al.* (1990) and Young *et al.* (1991) (figure 1). In this model, each input discharge produces an instantaneous increase in the postsynaptic voltage which then decays exponentially with a fixed time constant. Inputs from different fibres that overlap in time are summed, and when the postsynaptic voltage reaches a fixed threshold value, an output spike is recorded. The postsynaptic voltage is then reset to zero, and a period of refractoriness is imposed during which no output spikes can be generated. The two most important parameters for this coincidence mechanism are (i) the amplitude of the individual inputs relative to the threshold, and (ii) the time constant determining the duration of each input.

Because the model AN inputs could have different

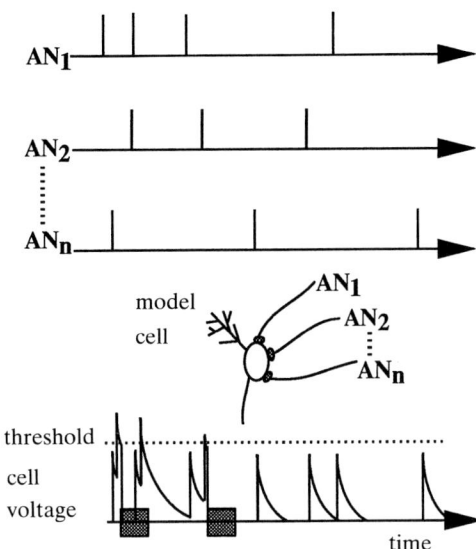

Figure 1. Schematic diagram of the AVCN cell model. At the top are responses of model AN fibres. A spike on any of the input fibres causes an exponentially decaying change in the postsynaptic cell voltage (lowest trace). When the voltage crosses the threshold value, an output spike is recorded for the model cell. After the occurrence of an output spike, the cell voltage is reset, and there follows a period of refractoriness (grey box) during which further spikes are ignored. The travelling wave delays for the different CFs are normalized before performing the coincidence detection.

CFs and thus different response latencies, it was necessary to adopt a rule for the relative neural delays that would determine the arrival times of input discharges converging on a postsynaptic cell. Based on the strong responses of cells with globular bushy response types to transient stimuli such as clicks, it was decided to 'align' the times of the inputs in response to the click by eliminating the differences between the traveling wave delays for inputs having different CFs.

3. RESULTS

Figure 2 shows the responses of a model cell to three Huffman sequences. These stimuli have a flat magnitude spectrum, but contain a phase transition of 2π that can be positioned at any frequency (F_t) and varied in slope (see Carney (1990*b*) for full description of the stimulus). The phase transition in the Huffman sequence provides a way to manipulate the relative discharge times of AN fibres with CFs near F_t; because individual fibres phase lock to components in the stimulus near their CF, the phase transition in the stimulus is represented in the phase-locked responses of the fibres. The model cell received inputs from 17 AN fibres with CFs distributed between 850 and 1150 Hz. Each input had an amplitude that was 0.8 times the threshold of the model cell, and the time constant for decay of each input was 0.05 ms. The model responses show a similar pattern of sensitivity to changes in the stimulus phase spectrum as that observed in physiological recordings from many cells, with the lowest threshold and the highest probability of response to the stimulus with the most gradual slope

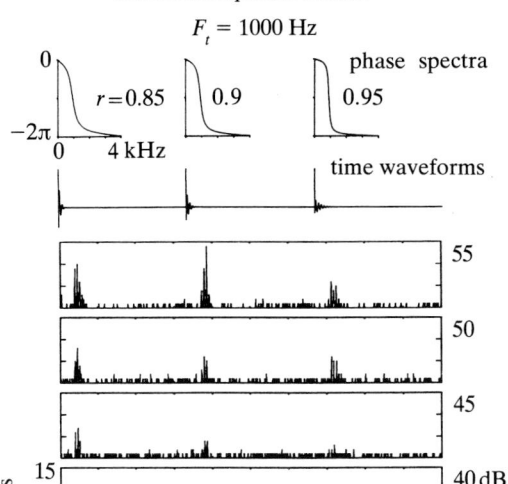

Figure 2. Responses of a model AVCN cell to three Huffman sequences (same format as in Carney (1990*b*)). At the top are the phase spectra and time waveforms for each of the three transient stimuli. The phase spectra have a transition centered at $F_t = 1000$ Hz. This transition increases in slope as the parameter r increases; the magnitude spectra for all three stimuli are flat (see Carney (1990*b*) for further description of stimuli.) Histograms of the responses of the model cell to these waveforms presented at four different sound pressure levels are shown. The model cell's threshold is lowest for the stimulus with the most gradual phase transition (left), and it responds to this stimulus with a higher probability than for the other stimuli as sound pressure is increased over this range (100 repetitions of the stimulus; 0.1 ms bins).

in the phase spectrum (Carney 1990*b*). Thus the coincidence mechanism acting on a group of inputs with distributed CFs can produce the type of sensitivity observed in the AVCN. If the amplitude of individual inputs was increased to near or at the threshold of the model cell, the sensitivity disappeared, as expected. Also, the sensitivity disappeared for a model cell receiving inputs having the same CF.

There are several qualitative differences between the model response and those recorded in the AVCN. In particular, the responses of the model are not as strongly locked into an onset-like response, the model cell has relatively high spontaneous rate, and the probability of firing of the model is lower than that of most globular bushy response types that were studied. Some of these differences might be reduced by further increasing the numbers of inputs to the postsynaptic model cell, and at the same time increasing the requirements for coincidence. However, some of the differences, particularly the difference in timing at the onset, may be due to limitations of the AN model used to provide the inputs. For simplicity, the AN inputs all had high spontaneous rates (Liberman 1991) and similar thresholds, all inputs produced postsynaptic responses with the same amplitude, and the postsynaptic cell's threshold remained constant over time. Models being developed that include the nonlinear features of the globular bushy cell's

membrane properties (Young *et al.* 1991) might improve some of these aspects of the responses. In addition, more accurate modelling might require the inclusion of other non-primary inputs, possibly including inhibitory inputs.

4. DISCUSSION

This study represents an initial test of the AVCN model's ability to reproduce the sensitivity to spatio-temporal patterns that was demonstrated physiologically. Coincidence detection in the AVCN might provide a mechanism for processing of the temporal information associated with responses to complex sounds. Different complex sounds evoke different spatiotemporal response patterns across the population of AN fibres, and thus a mechanism that is sensitive to differences in these patterns may be important in the processing of information that allows discrimination of complex sounds.

A coincidence mechanism acting upon convergent afferents with different CFs, as modelled here, may explain several response features of globular bushy cells that have been reported previously. A coincidence mechanism is consistent with the generally low spontaneous rate of this population (Smith *et al.* 1991). Also, coincidence detection could produce both the high synchronization coefficients that have been described for low-CF globular bushy cells and possibly small spherical bushy cells in response to low-frequency tones (Carney 1990*b*; Yin *et al.* 1988) and for high-CF globular bushy cells in response to low-frequency tones (Smith *et al.* 1991). Furthermore, this mechanism might explain the relatively high synchronization of these cells to the envelope of amplitude-modulated stimuli (Frisina 1990; Wang 1991). Coincidence detection, together with convergence of inputs of different CFs, might explain the relatively low sustained discharge rate following the strong onset response for many globular bushy cells to tone bursts at CF (Smith & Rhode 1987). Finally, the effect of convergence of different CFs on the sharpness of tuning need not be to broaden the tuning curve, which would be a result of convergence of different CFs without a coincidence requirement (see Carney (1990*b*) for discussion).

It is interesting to consider the possible functional significance of a coincidence mechanism in the AVCN. The pattern of sensitivity observed in the physiological recordings from several of these cells (Carney 1990*b*) suggests the hypothesis that this mechanism could enhance the signal-to-noise ratio of the temporal information that is transmitted from the cochlear nucleus to the binaural brainstem nuclei. Such an enhancement would be produced by the fact that the temporal responses to signal components (or spectral peaks) in complex sounds would be more highly correlated across fibres of different CFs than would be the responses to noise components of the stimulus. Thus the signal-related temporal information would pass more readily through the convergence and coincidence detection mechanism than would the noise-related temporal responses.

REFERENCES

Carney, L.H. 1990*a* A model for responses of auditory nerve fibers in cat. *Assoc. Res. Otolaryngol.* **13**, 140, (Abstract.)

Carney, L.H. 1990*b* Sensitivities of cells in the anteroventral cochlear nucleus of cat to spatio-temporal discharge patterns across primary afferents. *J. Neurophysiol.* **64**, 437–456.

Carney, L.H. 1992 A model for the responses of low-frequency auditory nerve fibers in the cat. (In preparation.)

Carney, L.H. & Yin, T.C.T. 1988 Temporal coding of resonances by low-frequency auditory nerve fibers: single fiber responses and a population model. *J. Neurophysiol.* **60**, 1653–1677.

Colburn, H.S., Han, Y-A. & Culotta, C.P. 1990 Coincidence model of MSO responses. *Hear. Res.* **49**, 335–346.

Frisina, R.D., Smith, R.L. & Chamberlain, S.C. 1990 Encoding of amplitude modulation in the gerbil cochlear nucleus: I. A hierarchy of enhancement. *Hear. Res.* **44**, 99–122.

Kiang, N.Y.S., Watanabe, T., Thomas, E.C. & Clark, C.F. 1965 *Discharge patterns of single fibers in the cat's auditory nerve.* Cambridge, Massachusetts: MIT Press.

Liberman, M.C. 1991 Central projections of auditory-nerve fibers of differing spontaneous rate. I. Anteroventral cochlear nucleus. *J. comp. Neurol.* **313**, 240–258.

Lorente de No, R. 1981 *The primary acoustic nuclei* New York: Raven.

Osen, K.K. 1970 Course and terminations of the primary afferents in the cochlear nuclei of the cat. An experimental anatomical study. *Archs ital. Biol.* **108**, 21–51.

Pfeiffer, R.R. 1966 Anteroventral cochlear nucleus: wave forms of extracellularly recorded spike potentials. *Science, Wash.* **154**, 667–668.

Rose, J.E., Brugge, J.F., Anderson, D.J. & Hind, J.E. 1967 Phase-locked responses to low-frequency tones in single auditory nerve fibers of the squirrel monkey. *J. Neurophysiol.* **30**, 769–793.

Smith, P.H., Joris, P.X., Carney, L.H. & Yin, T.C.T. 1991 Projections of physiologically characterized globular bushy cell axons from the cochlear nucleus of the cat. *J. comp. Neurol.* **304**, 387–407.

Smith, P.H. & Rhode, W.S. 1987 Characterization of HRP-labeled globular bushy cells in the cat anteroventral cochlear nucleus. *J. comp. Neurol.* **266**, 360–375.

Wang, X. 1991 Neural encoding of single-formant stimuli in the auditory-nerve and anteroventral cochlear nucleus of the cat. Ph.D. thesis, Johns Hopkins University, Maryland.

Wu, S.H. & Oertel, D. 1984 Intracellular injection with horseradish peroxidase of physiologically characterized stellate and bushy cells in slices of mouse anteroventral cochlear nucleus. *J. Neuroscience* **4**, 1577–1588.

Yin, T.C.T., Carney, L.H., Joris, P.X. & Smith, P.H. 1988 Enhancement of phase-locking to low frequency tones in some cells of the ventral cochlear nucleus of the cat. *Soc. Neurosci. Abstr.* **18**, 648.

Young, E.D., Rothman, J.S. & Manis, P.B. 1992 Regulatory of discharge constrains models of ventral cochlear nucleus bushy cells. In *The mammalian cochlear nuclei: organization and function* (ed. M. Merchen & J. Juiz). New York: Plenum Publishing Corporation. (In the press.)

Neural organization and responses to complex stimuli in the dorsal cochlear nucleus

ERIC D. YOUNG[1], GEORGE A. SPIROU[1]†, JOHN J. RICE[1] AND HERBERT F. VOIGT[2]

[1]Department of Biomedical Engineering and Center for Hearing Sciences, The Johns Hopkins University, Baltimore, Maryland 21205, U.S.A.
[2]Department of Biomedical Engineering, Boston University, Boston, Massachusetts 02215, U.S.A.

SUMMARY

The dorsal division of the cochlear nucleus (DCN) is the most complex of its subdivisions in terms of both anatomical organization and physiological response types. Hypotheses about the functional role of the DCN in hearing are as yet primitive, in part because the organizational complexity of the DCN has made development of a comprehensive and predictive model of its input–output processing difficult. The responses of DCN cells to complex stimuli, especially filtered noise, are interesting because they demonstrate properties that cannot be predicted, without further assumptions, from responses to narrow band stimuli, such as tones. In this paper, we discuss the functional organization of the DCN, i.e. the morphological organization of synaptic connections within the nucleus and the nature of synaptic interactions between its cells. We then discuss the responses of DCN principal cells to filtered noise stimuli that model the spectral sound localization cues produced by the pinna. These data imply that the DCN plays a role in interpreting sound localization cues; supporting evidence for such a role is discussed.

1. SYNAPTIC ORGANIZATION OF THE DCN

The dorsal cochlear nucleus (DCN) is a laminated structure which occupies the caudal and dorsal portion of the cochlear nuclear complex (Lorente de Nó 1981; Osen 1969). In the cat, the lamination of the DCN can be described in terms of the structure of one of the principal cell types of the nucleus, the fusiform or pyramidal cell (Blackstad et al. 1984); these cells have apical and basal dendritic trees. Their apical dendritic trees extend into the superficial molecular layer of the DCN. The cell bodies of the fusiform cells define the second layer of the nucleus and their basal dendrites extend into the deep DCN. The deep DCN contains the other DCN principal cell type, the giant cell, whose dendrites are mainly confined to the deep layers of the nucleus (Ryugo & Willard 1985). Axons of fusiform and giant cells project to the contralateral inferior colliculus (Adams 1979; Ryugo & Willard 1985). The DCN also contains small and mediun sized interneurons, which are distributed differently across the DCN laminae (Lorente de Nó 1981; Mugnaini et al. 1980; Osen et al. 1990). Some aspects of the relationships of the interneurons and principal cells in DCN are illustrated in the wiring diagram shown in figure 1. This figure does not show the lamination of the nucleus; instead, it shows the organization of connec-

tions as they are related to the major fibre systems that provide excitatory inputs to DCN neurons.

Auditory nerve (AN) fibres enter the DCN from the underlying posteroventral cochlear nucleus (PVCN) and project in an orderly fashion in the deep layers, preserving the monotonic sequence of best frequencies (BFS) generated in the cochlea (Osen 1970). As a result, the DCN is tonotopically organized, consisting of a stack of isofrequency sheets oriented parasaggitally (Spirou et al. 1989). The dendritic trees of both fusiform and giant cells are flattened and oriented in parallel with the isofrequency sheets (Blackstad et al. 1984; Ryugo & Willard 1985). Figure 1 shows two isofrequency sheets running from left to right. These sheets contain both fusiform (F) and giant (G) cells and are defined by the projections of AN fibres of two best frequencies (BF_1 and BF_2). Another source of excitatory input to the deep DCN comes from axon collaterals of multipolar cells in the PVCN (Oertel et al. 1990), and these also seem to project to isofrequency sheets.

The second major excitatory fibre system in the DCN is formed by the axons of granule cells which run as parallel fibres in the molecular layer (Mugnaini et al. 1980). Figure 1 shows that granule cell axons run perpendicular to the isofrequency sheets defined by deep layer tonotopic inputs. Granule cell axons make excitatory contacts on the apical dendritic trees of fusiform cells (Manis 1989); presumably, there are no granule cell inputs to the giant cells although there is little information on this issue. Granule cell axons also

† Present address: Department of Otolaryngology/Head and Neck Surgery, West Virginia University, Morgantown, West Virginia 26506 U.S.A.

Phil. Trans. R. Soc. Lond. B (1992) **336**, 407–413
Printed in Great Britain

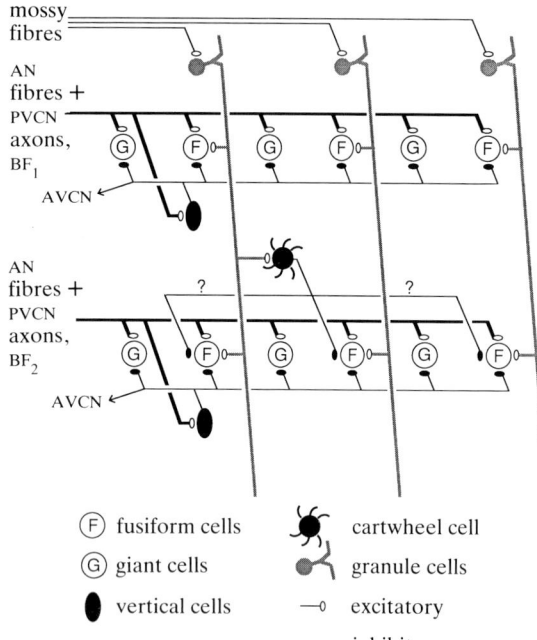

Figure 1. Partial wiring diagram of the DCN. Cell types are represented by symbols as shown in the legend. Terminals are shown as excitatory (unfilled) or inhibitory (filled). Each isofrequency sheet is represented by a horizontal row of principal cells (F and G); two best frequencies (BF_1 and BF_2) are shown. The approximate laminar relationships of axons can be inferred from occlusions where axons cross.

terminate on three varieties of interneurones in the superficial DCN (Osen *et al.* 1990). Only one of these, the cartwheel cell, is shown in figure 1.

The mossy fibre inputs to granule cells are apparently from diverse sources which have not been identified. One source which is of interest later in this paper is a projection from the dorsal column nuclei of the somatosensory system (Itoh *et al.* 1987; Weinberg & Rustioni 1987). Because the granule cell axons run orthogonal to isofrequency sheets, and therefore contact DCN cells across a range of BFs, the granule cell system seems not to conform to traditional ideas about auditory signal processing, which depend on separation of a signal into its component frequencies. Instead, granule cell axons are positioned to coordinate processing across frequencies, perhaps by selectivity activating different sets of principal cells or by regulating overall DCN excitability.

Vertical cells are inhibitory interneurones found in the deep layers; they probably receive excitatory inputs from the fibres in deep DCN and their axons innervate fusiform cells and presumably also giant cells (Lorente de Nó 1981; Oertel & Wu 1989; Osen *et al.* 1990). Vertical cell axons terminate profusely within what appears to be an isofrequency sheet in the DCN and then project to the anteroventral division of the cochlear nucleus (AVCN), where their axons make additional contacts, also within the same isofrequency sheet (Wickesberg & Oertel 1990). Vertical cells contain inhibitory amino acids (Osen *et al.* 1990) and have been shown to inhibit AVCN principal cells (Wickesberg & Oertel 1990) and DCN principal cells (see below; Voigt & Young 1980, 1990).

Cartwheel cells of the superficial DCN are a second inhibitory interneuron; they receive excitatory input from granule cells but probably not from AN fibres (Osen *et al.* 1990; Berrebi & Mugnaini 1991; Oertel & Wu 1989). Cartwheel cell axons seem to project to an isofrequency sheet; they terminate on fusiform cell somata and basal dendritic trees. Although they are positioned to terminate on other DCN cell types, such as vertical cells, it is not known whether they do so. Less is known about two additional types of inhibitory interneurons in the superficial layer, the Golgi and stellate cells.

Figure 1 suggests that the DCN is organized into tightly bound isofrequency sheets with cross-sheet coordination only via the axons of granule cells. In fact the extent of interaction and overlap of cell processes from adjacent isofrequency sheets is likely to be larger than is implied in the figure. Although the vertical, giant and fusiform cells have flattened dendritic trees oriented parallel to the isofrequency sheets (Blackstad *et al.* 1984; Osen *et al.* 1990; Ryugo & Willard 1985), the degree of overlap of terminal fields of the various DCN axonal and dendritic systems is not known from anatomical studies. For example, it is not clear whether the axonal terminal field of a vertical cell lies in the same isofrequency sheet as its dendritic tree. Information about the interactions of DCN neurons of different types and BFs has been obtained by cross-correlation analysis of multiple simultaneous single unit recordings (Voigt & Young 1988, 1990). In the next sections, this evidence is reviewed, after the response types of DCN neurons are defined.

2. RESPONSE PROPERTIES OF DCN NEURONS

Figure 2*a*, *b* shows response maps of two DCN units. These maps are plots of excitatory and inhibitory response areas for tone bursts of various frequencies and sound levels. Cochlear nucleus units can be classified into five groups, types I through V, based on the relative amount of excitation and inhibition in their response maps (Evans & Nelson 1973; Young 1984). Most units in the DCN fall into types II, III, or IV; in this paper, we discuss only types II and IV. Type II responses are recorded from vertical cell interneurons and type IV responses are recorded from principal cells, both fusiform and giant cells (Young 1980).

Type II units show zero spontaneous activity, a narrowly tuned excitatory area (figure 2*a*), and weak or no response to broadband stimuli (Young & Voigt 1982). The noise response is unusual, because type II units are the only cochlear nucleus neurons that respond so weakly to broadband noise. The most straightforward explanation for their weak noise responses is that type II units have large inhibitory sidebands that are not revealed in response maps like figure 2*a* because of the lack of spontaneous activity. When a low-level BF tone is used to produce background activity in a type II unit and a response map is constructed with a second tone, the result shows a strong inhibitory sideband above BF and a weaker one

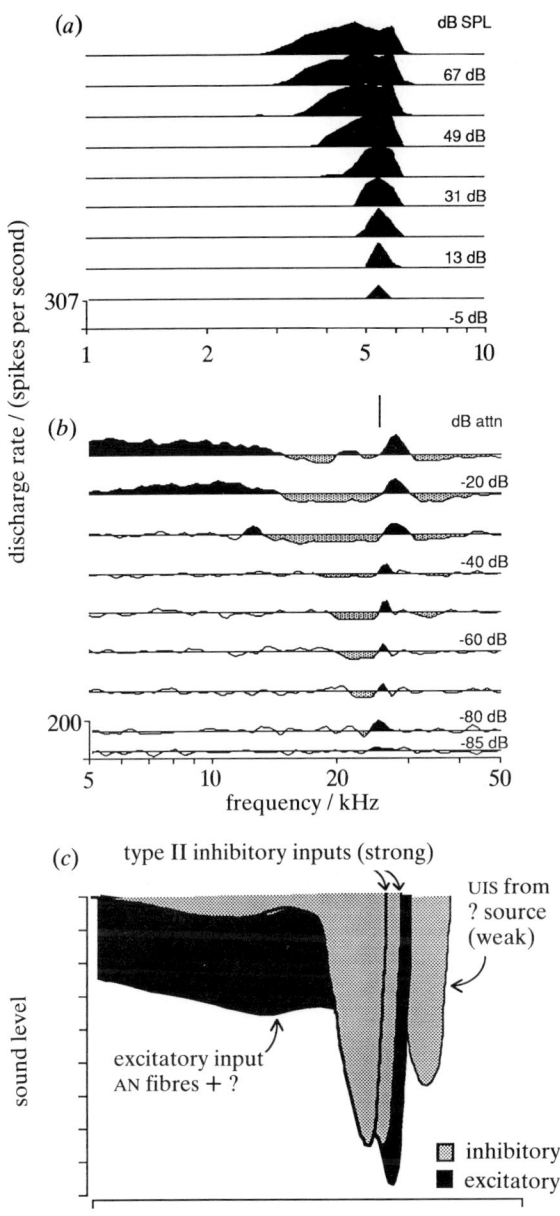

Figure 2. (*a*) Response map of DCN type II unit; map consists of ten plots, each showing discharge rate versus stimulus frequency for a sequence of 200 ms tone bursts (one per second) presented at one sound level. Sound levels given at right in dB SPL. Horizontal lines show spontaneous rate in each plot (0 spikes per second for this unit). Scale for rate plots given at lower left. Excitatory response regions (increase in discharge rate above spontaneous) filled with black. (*b*) Response map of DCN type IV unit; map constructed as in (*a*) except that this unit has spontaneous activity (≈ 50 spikes per second), so that inhibitory response areas can be seen (shaded regions where rate drops below spontaneous). Because of the way data were taken, sound levels are shown at right in dB attenuation; 0 dB is ≈ 100 dB SPL, but varies with frequency. Vertical line at top of response map is at BF. (*c*) Schematic model for generation of type IV response maps. Reproduced from Spirou & Young (1991) with permission.

below (Spirou & Young 1991). Whether these inhibitory areas are sufficient to account quantitatively for type II responses to broadband stimuli is not yet clear.

Type IV units are characterized by the response

map features shown in figure 2*b*. Usually the first response observed at low levels is excitatory, as at -80 and -85 dB. At higher levels, there is an inhibitory area (central inhibitory area or CIA) which is usually centred below BF, but which extends upward in frequency to include BF (Spirou & Young 1991). Because the CIA encroaches on BF, type IV units have a non-monotonic rate response to BF tone bursts: the response is excitatory over the first ≈ 30 dB above threshold and then becomes inhibitory. In most type IV units, there is an excitatory edge at the upper frequency limit of the CIA which separates the CIA from an upper inhibitory sideband at higher frequencies. At low frequencies, type IV units usually have a large excitatory region.

Figure 2*c* shows a model for the generation of type IV response maps based on the assumption that the CIA is produced by inhibitory inputs from type II units; the evidence for this connection has been discussed elsewhere (Young & Brownell 1976; Voigt & Young 1980, 1990) and is reviewed in the next section. The model consists of three components: (i) an excitatory input which has the characteristics of an AN tuning curve (black region), but which may be generated by both AN and PVCN inputs (Oertel *et al.* 1990); (ii) inhibitory inputs from type II units, shown by the two shaded tuning curves superimposed on the tunings curve of the excitatory input; and (iii) a second inhibitory input from an unknown source which produces the upper inhibitory sideband (UIS).

The type II inhibitory input is assumed to be strong, so that when a tone burst is within the tuning curve of the inhibitory input, the type IV unit is inhibited, producing the CIA. The BFs of the inhibitory inputs are shown slightly below the unit's excitatory BF based on two lines of evidence. First, the BFs of CIAS are generally somewhat below excitatory BFs in type IV units (Spirou & Young 1991). Second, the BFs of type II units that show inhibitory intractions with type IV units are more likely to be below than above the excitatory BFs of the type IV units (Voigt & Young 1990). The thresholds of the type II inputs are shown elevated by about 10 dB because type II units tend to have higher thresholds than type IV units (Young & Brownell 1976). These features account for the excitatory areas in type IV response maps near BF threshold and at the upper frequency edge of the CIA. The low frequency excitatory areas seen in type IV response maps result from the difference in low frequency thresholds between AN fibres (40–60 dB above threshold; Kiang 1984) and type II units (70–90 dB above threshold; figure 2*a* and Young & Voigt (1982)). There is variability in type IV response maps (Spirou & Young 1991), some of which can be accounted for by changes in the relative positions of the type II and excitatory inputs' BFs.

3. FUNCTIONAL ORGANIZATION OF THE DCN REVEALED BY CROSS CORRELATION

A straightforward way of studying the functional organization of a nucleus is to look for signs of correlated discharge among its neurons. If neurons are synaptically connected or receive common sources of input, then their spike trains should be correlated. Two types of correlation that are seen between pairs of neurons in the DCN are illustrated in figure 3a, b. Figure 3a shows a cross correlogram between a type II and a type IV unit. The type II unit is the reference, so the correlogram shows the average discharge rate of the type IV relative to spike times in the type II. Immediately to the right of the origin, there is a decrease in the correlogram implying that the probability of type IV spikes is reduced transiently after

type II spikes. This inhibitory trough (IT) is shaded in figure 3a. An IT is expected in the correlogram of a pair of units connected by a monosynaptic inhibitory synapse (Moore *et al.* 1970). The correlogram in figure 3b was computed from two type IV spike trains. The shaded increase in correlation near the origin, called a central mound (CM), is a sign of shared input; a CM can be produced by any effect that tends to synchronize the discharges of two neurons, including a common source of excitatory or inhibitory input (Moore *et al.* 1970).

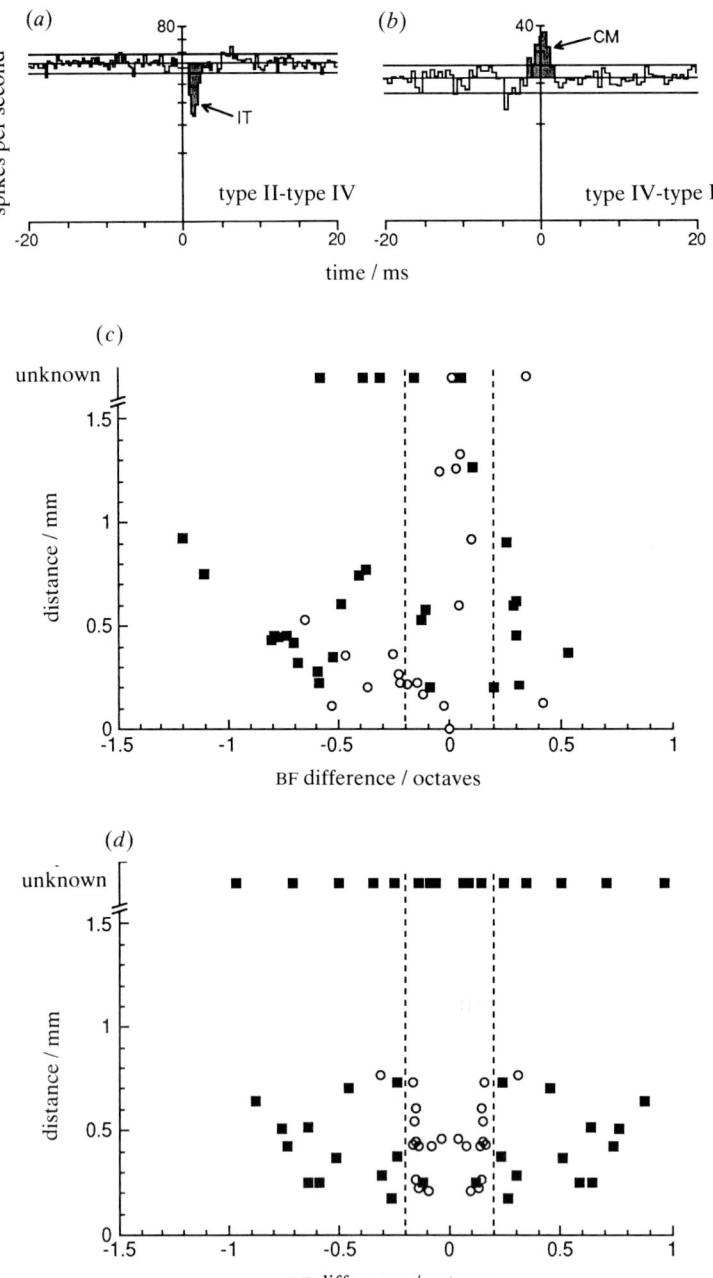

Figure 3. (a) Cross-correlogram of spike trains of simultaneously isolated DCN type II and type IV units. Horizontal lines show expected correlogram value for independently discharging neurons (centre) and ± 2 standard deviation confidence limits. Ordinate scale is square root to stabilize correlogram variance. (b) Same as (a) for type IV–type IV pair. (c) Difference in octaves between type II BF and type IV BF (abscissa) and distance apart in histological preparations (ordinate) for pairs showing and not showing an IT in their cross-correlograms. Data points at top of the figure come from pairs in which histological measure of distance is not available. (d) Same for type IV–type IV pairs showing and not showing CM correlation. Each data point plotted twice along the abscissa, at plus and minus the BF difference, because these pairs are symmetric. Redrawn from Voigt & Young (1988, 1990) with permission.

The data in figure 3 were obtained in experiments in which two electrodes were used to isolate units (Voigt & Young 1988, 1990). By looking for correlated pairs with the electrodes placed at various distances apart and in various orientations, it is possible to map out the spread of type II inhibitory connections onto type IV units and to look for the extent and spread of shared input to DCN principal cells. Figure 3c, d shows results on the occurrence of correlation as a function of the difference in BFs of the two neurons of the pair (abscissae) and as a function of the distance apart of the two neurons (ordinates). In each plot, filled symbols represent pairs with no sign of correlation under any stimulus condition tested and open symbols show pairs that were correlated with an IT (type II–IV pairs in figure 3c) or a CM (type IV–IV pairs in figure 3d). Difference in BF, i.e. position along the abscissa, is a measure of the distance apart of the units of a pair along the tonotopic axis. The distance measure on the ordinate includes both separation along and perpendicular to the tonotopic axis.

Figure 3c shows that type II inhibitory connections spread widely within isofrequency sheets near their cell bodies. The dashed lines show BF differences of ± 0.2 octaves. Within this frequency range, 12 out of 20 (60%) of pairs are correlated; moreover, correlation is observed at all distances examined (up to 1.325 mm, more than half way across the nucleus). Correlation is less commonly observed when the type II BF is 0.2–0.6 octaves above the type IV BF (2 out of 8, 25%) or 0.2–0.6 octaves below the type IV BF (6 out of 14, 43%). The limit on the distance that inhibitory connections spread is actually more stringent than is implied by these statistics. A dimensionless measure of IT amplitude was defined by Voigt & Young (1990); this measure decreases at a rate of only -0.2 per millimetre within an isofrequency sheet (i.e. for pairs with BFs within ± 0.3 octave of each other). By contrast, it decreases at a rate of -4.32 per millimetre along the tonotopic axis, i.e. when BF differences are converted to distances in millimetres using the tonotopic map of the DCN (Spirou et al. 1989). Finally, type II–type IV pairs are more likely to be correlated when the type II BF is below the type IV BF, although the difference is not statistically significant. This tendency is consistent with the fact that CIAs of type IV units generally have BFs somewhat below the excitatory BF (Spirou & Young 1991) and suggests that type II axons should spread to isofrequency sheets somewhat higher in frequency than those containing their dendrites.

Figure 3d shows the cumulative effect of all sources of input on type IV units in the form of the distribution of CM occurrence. Pairs are correlated essentially only within an isofrequency sheet. Eleven out of 16 pairs (69%) with BFs within ± 0.2 octave show a CM, whereas only one pair with a larger BF difference does so. It is instructive to compare the BF differences over which CMs are observed with an anatomical estimate of the width of an isofrequency sheet. The average thickness, along the tonotopic axis, of fusiform cell dendritic fields is 80 (basal) and 115 (apical) μm (Blackstad et al. 1984). The largest BF difference over which correlation was observed between a pair of type IV units in the fusiform cell layer is 0.138 octaves and the smallest BF difference at which an uncorrelated pair was observed is 0.239 octaves. These BF differences correspond to 97 and 167 μm, using the tonotopic map of the DCN (Spirou et al. 1989; see Voigt & Young, 1988). Thus, type IV units are correlated with other type IV units only when their dendrites are adjacent or overlap, i.e. within the same or adjacent isofrequency sheets.

The conclusion of the previous paragraph is consistent with the wiring diagram of figure 1, in that most axons in the DCN run parallel to isofrequency sheets. The major exception are granule cell axons. Thus, the results in figure 3d imply that granule cell inputs do not produce spike discharges in type IV units which are strongly correlated over a timescale of milliseconds, i.e. like the CM in figure 3b. The effects of granule cells might be weak or sparse, so that they are not seen in cross-correlation analysis, or they might involve long-term modulation of principal cell excitability.

There is no distinction between fusiform and giant cells in terms of the functional organization revealed by cross-correlation analysis. Both cell types show IT correlation with type II units (Voigt & Young 1990) and CMS are observed between pairs of fusiform cells, between pairs of giant cells, and between mixed pairs (Voigt & Young 1988). Thus the organization of both principal cell types into a common module, implied by figure 1, is supported by the cross-correlation evidence. The major difference between fusiform and giant cells appears to be that fusiform cells are and giant cells are not innervated by the granule-cell associated circuitry in superficial DCN. Apparently, the methods which have been used to study DCN principal cells are insensitive to the effects of the inputs to fusiform cells from the superficial DCN.

4. RESPONSES OF DCN PRINCIPAL CELLS TO COMPLEX STIMULI

DCN principal cells show excitatory responses to broadband stimuli, which is surprising given their generally inhibitory responses to tones (Young & Brownell 1976). This result can be explained by the characteristics of type II units, which respond strongly to tones, but weakly to broadband noise. These type II properties force type IV units to be inhibited by tones, but allow excitation by noise. When DCN principal cells respond to band-reject filtered noise, they show an additional interesting property, a sharp sensitivity to spectral nulls or notches in the stimulus spectrum (Evans 1977; Spirou & Young 1991).

In the cat, narrow spectral notches are produced naturally by the filtering properties of the pinna (Musicant et al. 1990; Rice et al. 1992). The frequencies at which these notches occur are dependent on sound source direction and therefore provide a sound localization cue. In particular, the cat pinna produces a prominent notch (called the first notch, FN) in the 5–20 kHz range for sounds originating in the frontal

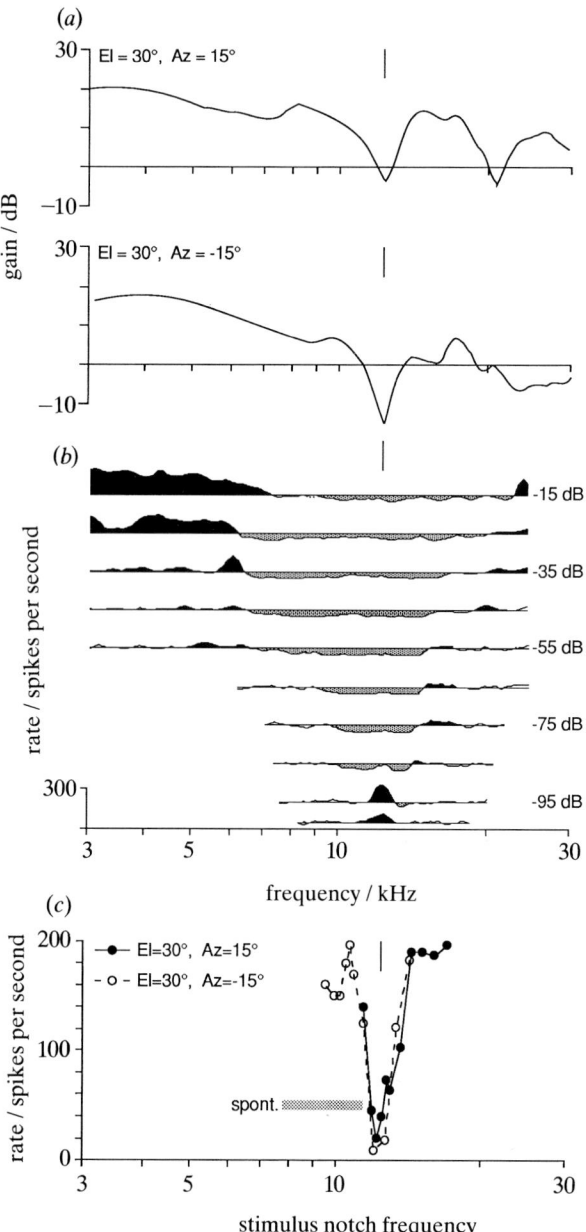

Figure 4. (*a*) Spectra at the eardrum when broad band noise is presented from two directions in space; spectra are shifted to align FN frequency with BF of unit in (*b*). (*b*) Response map of DCN type IV unit. (*c*) Discharge rate of unit as function of FN frequency for stimuli in (*a*). Fuzzy line is spontaneous rate.

stimuli in figure 4*a*. Broadband noises were filtered to have the spectra in figure 4*a* and were then presented as 200 ms bursts. In order to generate a battery of stimuli, the sampling frequency of the D/A converter was changed, which changes the frequencies of the FNs proportionally. The spectra are drawn in figure 4*a* for a sampling frequency which places the FNs at the unit's BF. Figure 4*c* shows the discharge rate of the unit as a function of the frequency of the FNs. The unit is typical of many DCN type IV units in that it undergoes a sharp change in response as a notch moves past its BF, with its response rate going from excitation to inhibition to excitation again.

The data in figure 4 show that DCN principal cells are exquisitely sensitive to spectral features of stimuli that carry sound localization information. This fact suggests that the DCN may be involved in early analysis of sound source location. Two additional findings are consistent with this hypothesis. First, Sutherland (1991) has reported that lesions of the output axons of the DCN produce deficits in localizing sound sources in elevation. Elevational sound localization, of course, depends entirely on spectral cues of the kind shown in figure 4*a*. Second, there is a projection from the dorsal column nuclei of the somatosensory system to the granule cell areas of the cochlear nucleus (Itoh *et al.* 1987; Weinberg & Rustioni 1987). The cells of origin of this projection are clustered in the lateral portion of the cuneate nucleus, where the representation of the back of the head and the pinna is located (Millar & Basbaum 1975). This correlation suggests that the somatosensory input to the DCN is providing information about the orientation of the cat's mobile pinnae and, by extension, that the DCN is integrating the pinna-position information with information about sound localization cues. Correcting for pinna orientation is essential to using spectral sound localization cues because the mapping between position in space and spectral cues changes when the pinna moves. Under this hypothesis, the orientation of the granule cell axons perpendicular to isofrequency sheets can be interpreted as providing non-auditory information about pinna position to coordinate the activity of DCN neurons of different BFs.

This work was supported by grants from the National Institutes of Health, the Deafness Research Foundation, and the National Science Foundation.

field. The frequency of the FN, by itself, is a sufficient cue for localizing sounds in front of the animal (Rice *et al.* 1992). Figure 4*a* shows shifted versions of the spectra at the eardrum produced when a broadband noise is presented from two locations in free field. The spectra are shifted for the reasons described below; the FNs of these spectra (vertical lines) actually occur at frequencies of 13.7 kHz (top spectrum) and 11.4 kHz (bottom spectrum). Clearly there are substantial differences in these spectra, including FN frequency, which the cat could use as sound localization cues.

Figure 4*b* shows the response map of a DCN type IV unit and figure 4*c* shows the sensitivity of this unit to the position along the frequency axis of the FNs of the

REFERENCES

Adams, J.C. 1979 Ascending projections to the inferior colliculus. *J. comp. Neurol.* **183**, 519–538.

Berrebi, A.S. & Mugnaini, E. 1991 Distribution and targets of the cartwheel cell axon in the dorsal cochlear nucleus of the guinea pig. *Anat. Embryol.* **183**, 427–454.

Blackstad, T.W., Osen, K.K. & Mugnaini, E. 1984 Pyramidal neurons of the dorsal cochlear nucleus: a Golgi and computer reconstruction study in cat. *Neuroscience*, **13**, 827–854.

Evans, E.F. & Nelson, P.G. 1973 The responses of single neurones in the cochlear nucleus of the cat as a function of their location and the anaesthetic state. *Expl Brain Res.* **17**, 402–427.

Evans, E.F. 1977 Frequency selectivity at high signal levels of single units in cochlear nerve and nucleus. In *Psychophysics and physiology of hearing* (ed. E. F. Evans & J. P. Wilson), pp. 185–196. London: Academic Press.

Itoh, K., Kamiya, H., Mitani, A., Yasui, Y., Takada, M. & Mizuno, N. 1987 Direct projections from the dorsal column nuclei and the spinal trigeminal nuclei to the cochlear nuclei in the cat. *Brain Res.* **400**, 145–150.

Kiang, N.Y.S. 1984 Peripheral neural processing of auditory information. In *Handbook of physiology—the nervous system III* (ed. J. M. Brookhart & V. B. Mountcastle), pp. 639–674. Bethesda: Am. Physiol. Soc.

Lorente de Nó, R. 1981 *The primary acoustic nuclei.* New York: Raven Press.

Manis, P.B. 1989 Responses to parallel fiber stimulation in the guinea pig dorsal cochlear nucleus in vitro. *J. Neurophysiol.* **61**, 149–161.

Millar, J. & Basbaum, A.I. 1975 Topography of the projection of the body surface of the cat to cuneate and gracile nuclei. *Expl Neurol.* **49**, 281–290.

Moore, G.P., Segundo, J.P., Perkel, D.H. & Levitan, H. 1970 Statistical signs of synaptic interaction in neurons. *Biophys. J.* **10**, 876–900.

Mugnaini, E., Warr, W.B. & Osen, K.K. 1980 Distribution and light microscopic features of granule cells in the cochlear nuclei of cat, rat, and mouse. *J. comp. Neurol.* **191**, 581–606.

Musicant, A.D., Chan, J.C.K. & Hind, J.E. 1990 Direction-dependent spectral properties of cat external ear: new data and cross-species comparisons. *J. acoust. Soc. Am.* **87**, 757–781.

Oertel, D. & Wu, S.H. 1989 Morphology and physiology of cells in slice preparations of the dorsal cochlear nucleus of mice. *J. comp. Neurol.* **283**, 228–247.

Oertel, D., Wu, S.H., Garb, M.W. & Dizack, C. 1990 Morphology and physiology of cells in slice preparations of the posteroventral cochlear nucleus of mice. *J. comp. Neurol.* **295**, 136–154.

Osen, K.K. 1969 Cytoarchitecture of the cochlear nuclei in the cat. *J. comp. Neurol.* **136**, 453–482.

Osen, K.K. 1970 Course and termination of the primary afferents in the cochlear nuclei of the cat. *Archs ital. Biol.* **108**, 21–51.

Osen, K.K., Ottersen, O.P. & Storm-Mathisen, J. 1990 Colocalization of glycine-like and GABA-like immunoreactivities: a semiquantitative study of individual neurons in the dorsal cochlear nucleus of cat. In *Glycine neurotransmission* (ed. O. P. Ottersen & J. Storm-Mathisen), pp. 417–452. Chichester: John Wiley and Sons.

Rice, J.J., May, B.J., Spirou, G.A. & Young, E.D. 1992 Pinna-based spectral cues for sound localization in the cat. *Hearing Res.* **58**, 132–152.

Ryugo, D.K. & Willard, F.H. 1985 The dorsal cochlear nucleus of the mouse: a light microscopic analysis of neurons that project to the inferior colliculus. *J. comp. Neurol.* **242**, 381–396.

Spirou, G.A., May, B. & Ryugo, D.K. 1989 3-Dimensional frequency mapping in the cat dorsal cochlear nucleus. *Soc. Neurosci. Abstr.* **19**, 744.

Spirou, G.A. & Young, E.D. 1991 Organization of dorsal cochlear nucleus type IV unit response maps and their relationship to activation by bandlimited noise. *J. Neurophysiol.* **66**, 1750–1768.

Sutherland, D.P. 1991 A role of the dorsal cochlear nucleus in the localization of elevated sound sources. *Abst. Assoc. Res. Otolaryngol.* **14**, 33.

Voigt, H.F. & Young, E.D. 1980 Evidence of inhibitory interactions between neurons in the dorsal cochlear nucleus. *J. Neurophysiol.* **44**, 76–96.

Voigt, H.F. & Young, E.D. 1988 Neural correlations in the dorsal cochlear nucleus: Pairs of units with similar response properties. *J. Neurophysiol.* **59**, 1014–1032.

Voigt, H.F. & Young, E.D. 1990 Cross-correlation analysis of inhibitory interactions in dorsal cochlear nucleus. *J. Neurophysiol.* **64**, 1590–1610.

Weinberg, R.J. & Rustioni, A. 1987 A cuneocochlear pathway in the rat. *Neuroscience* **20**, 209–219.

Wickesberg, R.E. & Oertel, D. 1990 Delayed, frequency-specific inhibition in the cochlear nuclei of mice: a mechanism for monaural echo suppression, *J. Neurosci.* **10**, 1762–1768.

Young, E.D. 1980 Identification of response properties of ascending axons from dorsal cochlear nucleus. *Brain Res.* **200**, 23–38.

Young, E.D. 1984 Response characteristics of neurons of the cochlear nuclei. In *Hearing science, recent advances* (ed. C. I. Berlin), pp. 423–460. San Diego: College-Hill Press.

Young, E.D. & Brownell, W.E. 1976 Responses to tones and noise of single cells in dorsal cochlear nucleus of unanesthetized cats. *J. Neurophysiol.* **39**, 282–300.

Young, E.D. & Voigt, H.F. 1982 Response properties of type II and type III units in dorsal cochlear nucleus. *Hearing Res.* **6**, 153–169.

Discussion

A. REES (*Department of Physiological Sciences, University of Newcastle upon Tyne, U.K.*). The cytoarchitecture of the dorsal cochlear nucleus in man is different to that of the cat. What implications might this have for the operation in man of a localizing mechanism analogous to the one Professor Young has suggested for the cat?

E. D. YOUNG. Although there are many aspects of the differences between cat and primate DCN that have not been worked out satisfactorily, it is clear from published descriptions that differences exist in the superficial layers of the nucleus, i.e. in the portions of the DCN associated with the granule cells and their associated interneurons. In apes and man, this system is small or missing. In our paper, we hypothesize that the role of the superficial granule-cell-related system in the cat is to integrate auditory information about spectral sound localization cues with somatosensory information about pinna position. If this hypothesis is correct, then animals without mobile pinnae, like apes and man, would not need the granule-cell-associated circuitry in superficial DCN. A comparative study correlating pinna mobility and DCN organization would be very interesting.

Binaural masking and sensitivity to interaural delay in the inferior colliculus

A. R. PALMER[1], A. REES[2] AND D. CAIRD[3]

[1] MRC Institute of Hearing Research, University of Nottingham, University Park,
Nottingham NG7 2RD, U.K.
[2] Department of Physiological Sciences, The Medical School, Framlington Place,
Newcastle upon Tyne, NE2 4HH, U.K.
[3] Zentrum der Physiologie, Theodor Stern Kai 7, 6000 Frankfurt am Main 70, F.R.G.

SUMMARY

The binaural masking level difference (BMLD) is a psychophysical effect whereby signals masked by a noise at one ear become unmasked by sounds reaching the other. BMLD effects are largest at low frequencies where they depend on signal phase, suggesting that part of the physiological mechanism responsible for the BMLD resides in cells that are sensitive to interaural time disparities.

We have investigated a physiological basis for unmasking in the responses of delay-sensitive cells in the central nucleus of the inferior colliculus in anaesthetized guinea pigs. The masking effects of a binaurally presented noise, as a function of the masker delay, were quantified by measuring the number of discharges synchronized to the signal, and by measuring the masked threshold. The noise level for masking was lowest at the best delay for the noise. The mean magnitude of the unmasking across our neural population was similar to the human psychophysical BMLD under the same signal and masker conditions.

1. INTRODUCTION

Most low-frequency cells in the cat inferior colliculus are sensitive to interaural time differences. In effect, these cells respond to interaural phase differences (IPDs) in low-frequency components of the signal (Kuwada & Yin 1983). A plot of spike-rate against interaural time delay is a cyclical function whose period is either the reciprocal of the stimulation frequency or, for noise stimulation, a frequency close to the cell's best frequency (Kuwada & Yin 1983; Yin et al. 1986). The delay producing the main peak in a plot of the spike rate versus the interaural delay for broadband signals is termed the best delay and it usually corresponds to a sound source position in the contralateral hemified.

Human psychophysical studies show that the masking of binaural signals by broadband noise depends on the interaural phase relationship of signal and masker. This was first demonstrated in 1948 (Licklider 1948; Hirsh 1948) and has since been extensively studied (see Durlach & Colburn 1978). If identical low-frequency tones and noises are presented to both ears in human subjects, improved detectability of the tone can be achieved by varying the phase of either the tone or noise in one ear. These unmasking effects, which have also been demonstrated in animals (Wakeford & Robinson 1974; Cranford 1975), are termed the binaural masking level difference (BMLD).

The association of some of these BMLD effects with lateralization (Jeffress & McFadden 1971), the restriction of large BMLD effects to low frequencies, and the dependency of the BMLD on parameters such as the phase of the signal (Jeffress et al. 1962), together suggest that part of the physiological mechanism responsible for the BMLD resides in cells with similar properties to IPD cells.

There were two aims in the present study, first to quantify delay sensitivity in the guinea pig and second to investigate its relation to the BMLD. First we measured the best delays of a large sample of neurons in the guinea pig to allow a detailed comparison of the cat data with those from another mammal with a smaller head. Second, unlike earlier studies (Langford 1984; Caird et al. 1989) where the signals replicated the simple psychophysical BMLD test, we optimized our signals for the delay sensitivity of individual cells by measuring the masking effect of the noise on a signal at its best delay. We also investigated noise masking effects on the cells' synchronization to the fundamental frequency of a vowel sound. This experimental design, in which speech sounds are presented with one delay and noise maskers with other delays, has elements in common with the use of binaural cues in the 'cocktail party' effect: the separation of two or more spatially segregated sound sources. More detailed accounts of the data described here may be found elsewhere (Palmer et al. 1990; Caird et al. 1991).

Phil. Trans. R. Soc. Lond. B (1992) **336**, 415–422
Printed in Great Britain

415

2. METHODS

The methods are described in detail elsewhere (Palmer *et al.* 1990; Caird *et al.* 1991) and are only summarized here. The experiments were performed on mature, pigmented guinea-pigs under anaesthesia.

The stimuli were delivered dichotically through sealed acoustic systems. The three stimuli used were, pure tones, broadband noise and synthetic vowel sounds. The onsets and offsets of the tone and noise stimuli were shaped with 5 ms rise-fall times. The speech signal was a segment of a steady-state approximation to the vowel /a/, with a fundamental frequency (F_0) of 100 Hz (see Palmer *et al.* (1986) for details). A pair of two-channel delay lines enabled the interaural delays of the signals and a noise masker to be varied independently between the two ears with a resolution of 1 μs.

The activity of single units in the inferior colliculus was recorded with tungsten microelectrodes positioned stereotaxically. A micro-computer recorded the time of occurrence of the extracellularly recorded action potentials. Histological reconstructions confirmed that the neurones were isolated in the central nucleus of the inferior colliculus.

The delay sensitivity of neurons with best frequencies (BFs) below 4.0 kHz was measured in response to 50 ms bursts of BF tone, noise, and vowel, presented with a repetition rate of 5 per second. The best delays were estimated visually or, when at least one cycle of response was available, using a single cycle vector strength analysis (see Kuwada & Yin 1983).

In the first masking test, the signal (a 50 ms burst of best-frequency tone or vowel) was set at its best delay and level and the masked threshold was measured as a function of the delay of a continuous noise masker. The masking was assessed either (i) subjectively using audio-visual criteria, or (ii) objectively using a computer-controlled adaptive procedure (Taylor & Creelman 1967) in which the discharge rate in the noise-alone interval was compared with that in the signal plus noise interval and the noise level was adjusted until a criterion difference (one spike more in the signal interval) was achieved on 75% of presentations.

In the second masking test we examined the effects of the interaural delay of a continuous noise masker on the response to a 500 ms segment of the synthetic vowel. The vowel was set at its best delay and level and presented at a repetition rate of 1 Hz. The level of masking noise was adjusted so that it just masked the response to the vowel when at the same delay. The delay of the noise was then varied in 100 μs steps over the range ±2000 μs. The cell's mean firing rate during the signal plus noise and noise-alone intervals were calculated as was the discharge rate synchronized to the F_0 of the vowel.

Human psychophysical masking was measured in three subjects listening over headphones to the signals used in the physiological experiments. The synthetic vowel /a/ was presented at 62 dB SPL with a 300 μs right ear lead and a continuous wideband noise was used to mask the vowel. A method of adjustment was used to estimate the masked theshold as a function of the interaural delay of the noise masker.

3. RESULTS

(a) *Response as a function of the interaural delay of best frequency tones and noise*

Delay sensitivity was measured in 171 neurons with best frequencies from 120 to 1800 Hz. Twenty-eight percent of these cells (48 out of 171) and all 72 cells tested with BFs between 1800 and 4000 Hz were not sensitive to the interaural delay of the BF tone.

We first measured the response to a single presentation of a stimulus (noise or tone) as a function of both delay and mean binaural intensity level. Having established the level which gave the greatest variation in response as a function of interaural time delay, we then measured the delay function at that optimal intensity with repeated presentations of each delay in random order.

As in other species, the discharge of most units to low-frequency tones fluctuates as the delay is varied. This is illustrated for a single unit in figure 1 (open circles). The period of the cycling in figure 1 is 1350 μs which is close to the reciprocal of the stimulation frequency (1/760 Hz = 1315 μs), which in this case was at the unit BF.

The filled squares in figure 1 show the delay function in response to wideband noise. For this cell, the best delays for noise and tones are the same (300 μs). However, unlike the responses to pure tones, the fluctuations in the noise delay functions become less pronounced at larger delay values. Indeed, in the example shown there is only one peak in the noise delay function.

There was a wide variation in the shapes of the noise delay curves from different cells. Generally, the peak of the delay curve was flanked by one or two

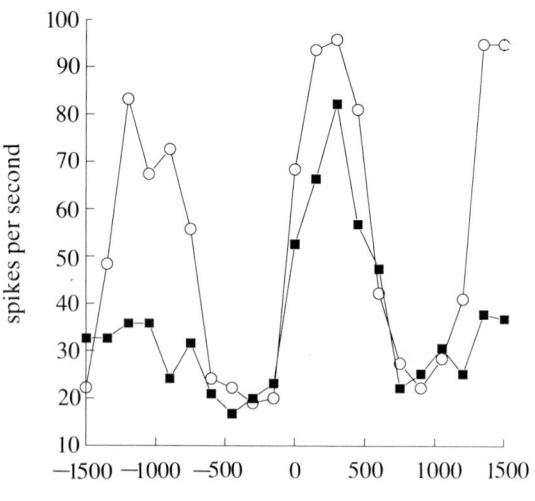

signal delay / μs contra lead

Figure 1. Response as a function of the interaural delay of best frequency tones (open circles: 760 Hz 55 dB SPL) and wideband noise (filled squares: 28 dB SPL spectrum level). (Replotted from figure 4a of Palmer *et al.* (1990).)

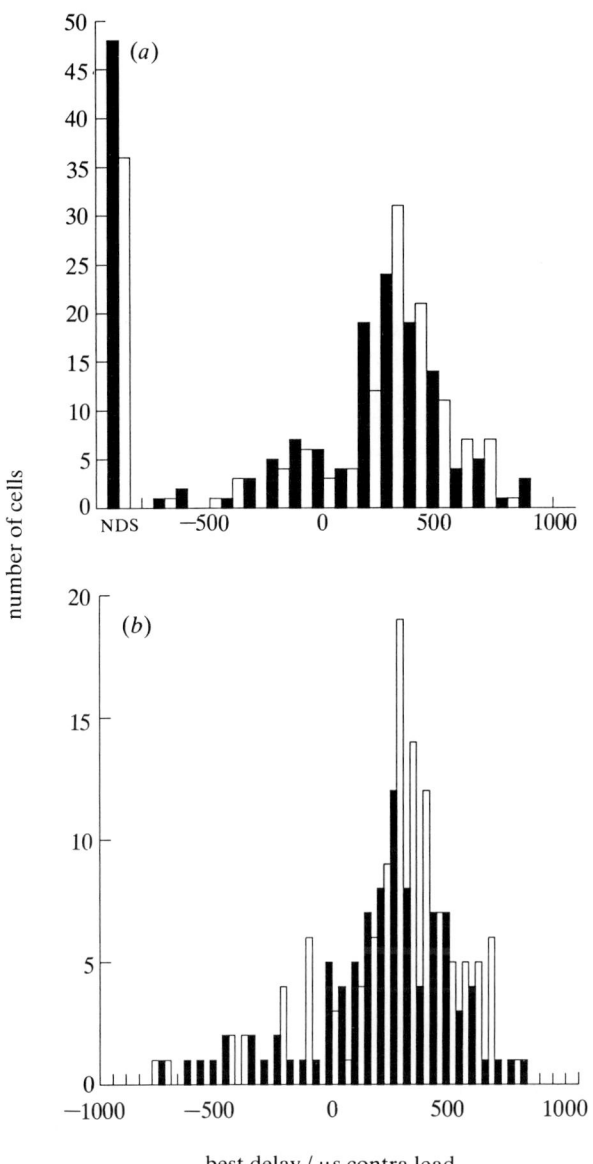

Figure 2. (*a*) Histogram of the number of neurons with different best delays. The filled bars are data obtained with best frequency tones and the unfilled bars are with wideband noise. Only cells with best frequencies below 1.8 kHz are included in the data obtained with pure tones. NDS indicates neurons not delay sensitive. (Replotted from figure 6, Palmer *et al* (1990).) (*b*) Comparison of the best delays to noise stimulation between the cat (filled bars: replotted from Yin *et al.* (1986)) and the guinea pig (unfilled bars).

troughs. However, although some cells showed sustained cycling, in others only a single peak was present and in yet others only a trough (see figure 5 in Palmer *et al.* (1990)). For these latter neurons the major effect of delaying the noise was to produce a reduction of the output near zero delay, without facilitation at other delays. For these cells (4%, 10 out of 237) we have used the position of the minimum in the delay function as our estimate of the best delay.

The range of best delays in our sample of neurons is shown as a histogram in figure 2*a*, for both BF tone and noise signals. Most of the neurons responded best when the stimulus at the contralateral ear occurred

100–400 μs before that to the ipsilateral ear, corresponding to sound sources in the contralateral hemifield. Figure 2*b* shows a direct comparison of our range of best delays (for noise stimulation) with data from the cat (replotted from Yin *et al.* (1986)). The distributions clearly overlap and are not statistically different ($t = 1.07$, $p > 0.05$; $F = 1.19$, $p > 0.05$).

(b) *Measurement of masking by comparisons of responses to signal plus masker with that to masker alone*

Masking in individual cells was quantified by comparing the response to signal plus noise with that to noise alone to determine the masked threshold. This method yields a measure that is directly analogous to the BMLD. We have completed these analyses on 76 single neurons (five audio-visually and 71 objectively), of which 43 gave a variation in masked threshold with the masker delay. Twenty eight units showed no delay dependent masking despite sensitivity to the interaural delay of the masking noise. These 28 units had no best frequency or delay sensitivity characteristics that set them apart from the other cells in our sample.

The level of noise required for masked threshold was lowest when the noise was at its best delay, and highest at the minimum of the noise delay function for 38 of the 43 cells showing delay dependent masking. Two examples are shown in figure 3. For each cell, the responses to the noise and signal presented alone are shown as a function of interaural delay in the upper panel, with the level of the masking noise required to achieve the masked threshold, at each noise delay, immediately below. The delay functions were used to select a delay which resulted in a strong response to the signal (shown by the arrows in figure 3); in most instances this was the best delay. The shapes of the masking functions are approximately the inverse of the noise delay functions (filled squares); the delay sensitivity to the signal or the nature of the signal are not relevant, since the signal merely establishes a level of driven activity against which the masking can be demonstrated. This can be seen in figure 3*b* where both the noise delay function and the masking function with a tone signal, cycle at 625 Hz, whereas the delay function to the tone alone cycles at the best frequency of 820 Hz (see Yin *et al.* (1987) for a description of this phenomenon).

Some IPD cells are characterized by delay-dependent inhibition rather than a peak due to facilitation, and the best delay of such cells is defined as the delay producing the strongest inhibition (see Palmer *et al.* 1990). For these cells, the response to the signal was inhibited by the noise and the lowest levels of noise required for masking were at delays corresponding to the minimum in the noise delay functions.

The correlation between the best delay to the noise and the minimum in the masking function is shown in figure 4. The deviation of the minimum in the masking function from the the noise best delay is plotted as a fraction of the noise response period. In most of the cells the minimum masked threshold

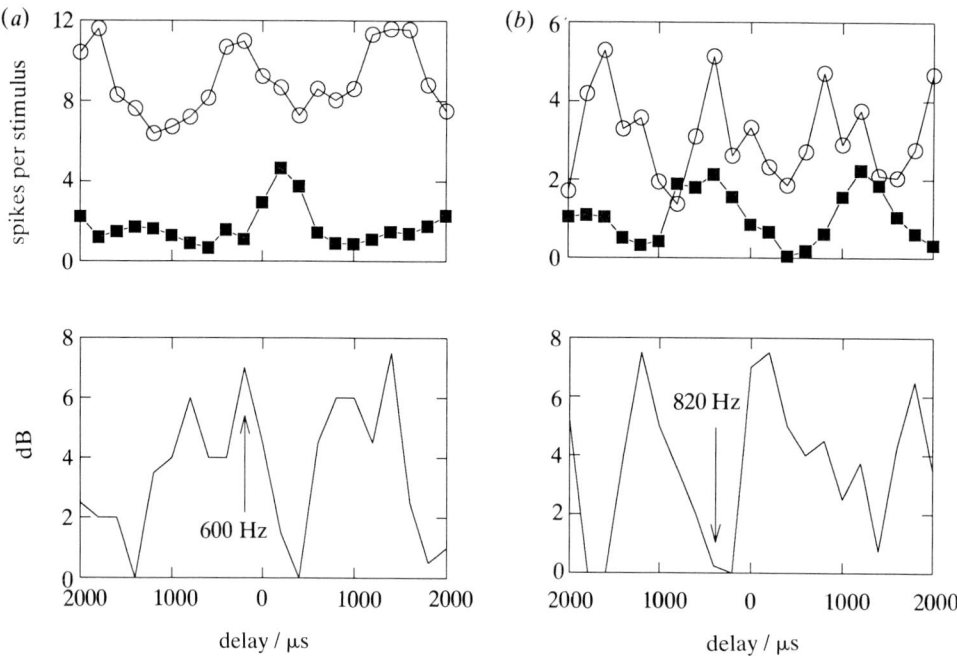

Figure 3. The upper panels show the delay functions from two different cells for noise (filled squares) and tone (open circles) stimuli. These delay functions were determined at the optimal sound level for each stimulus. The lower panels show the masked thresholds as a function of the interaural delay of the noise masker for the same cells. The labelled arrows indicate the signal interaural delay. (*a*) Data from a cell with BF of 600 Hz. The signal was a BF tone signal at 60 dB SPL. The zero reference level for the masked thresholds was 22 dB SPL spectrum level. (*b*) Data from a cell with BF of 820 Hz. The tone signal was presented at the best delay at 62 dB SPL. The zero reference level for the masked thresholds was 30 dB SPL spectrum level. (Replotted from figure 1, Caird *et al.* (1991).)

occurred when the noise was at or near (± 10%) its best delay.

For the cells that showed delay-dependent masking, the difference in noise level between the minimum and maximum masked thresholds varied between 5

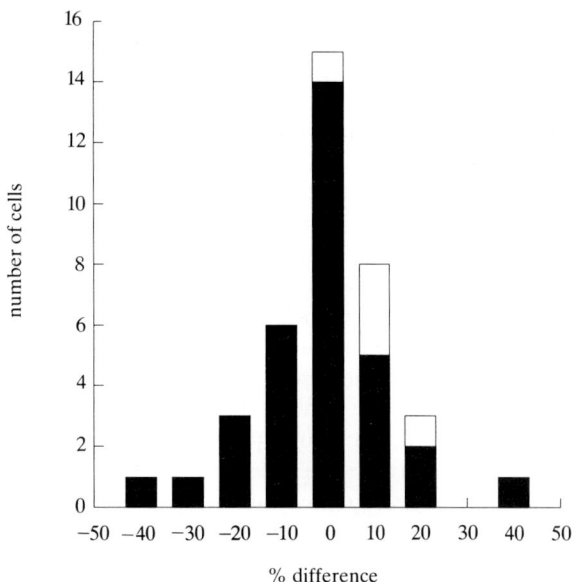

Figure 4. Histogram of the differences between the best delay of the cell to the masking noise and the noise delay producing the minimum masked threshold level. These differences have been expressed in terms of percent of the response period to the noise. The cells with inhibitory best delays are shown separately (clear bars). (Replotted from figure 5*b*, Caird *et al.* (1991).)

and 36 dB as shown in figure 5. To emphasize the similarities in cell responses we have eliminated the differences in best frequency and best delay by scaling the delays to fractions of the period of the noise delay function (± 100%) and shifting the functions to coincide at the noise best delay (0%). The noise level at which the masking curves are at the minimum is redefined as 0 dB. A small number of cells exhibited large variations in masked threshold and for these (dotted lines) the second ordinate has been used. The mean population responses were calculated at each 10% normalized delay inteval and are shown to the right (figure 5*b*, *d*). The peak-to-trough masking level differences in these mean curves are 6.9 dB and 6.8 dB for vowel and tone respectively. The maximum levels of noise to achieve masking are reached at approximately ± 50% and only a suggestion of cycling can be seen in the mean functions (figure 5*b*, *d*); cycling is seen in the masking functions of some individual neurons (figure 5*a*, *c* and figure 3*a*, *b*). The open symbols in figures 5*b*, *d* show psychophysical measures of binaural masking obtained under conditions comparable to those used to obtain the physiological data. The psychophysical data have been scaled to a single period of the stimulus (or of the first formant for the vowel) and shifted to coincide with the physiological data at the curve minimum.

(c) Measurements of masking within the signal interval

The measurements of masking described above do not provide information about the mechanisms under-

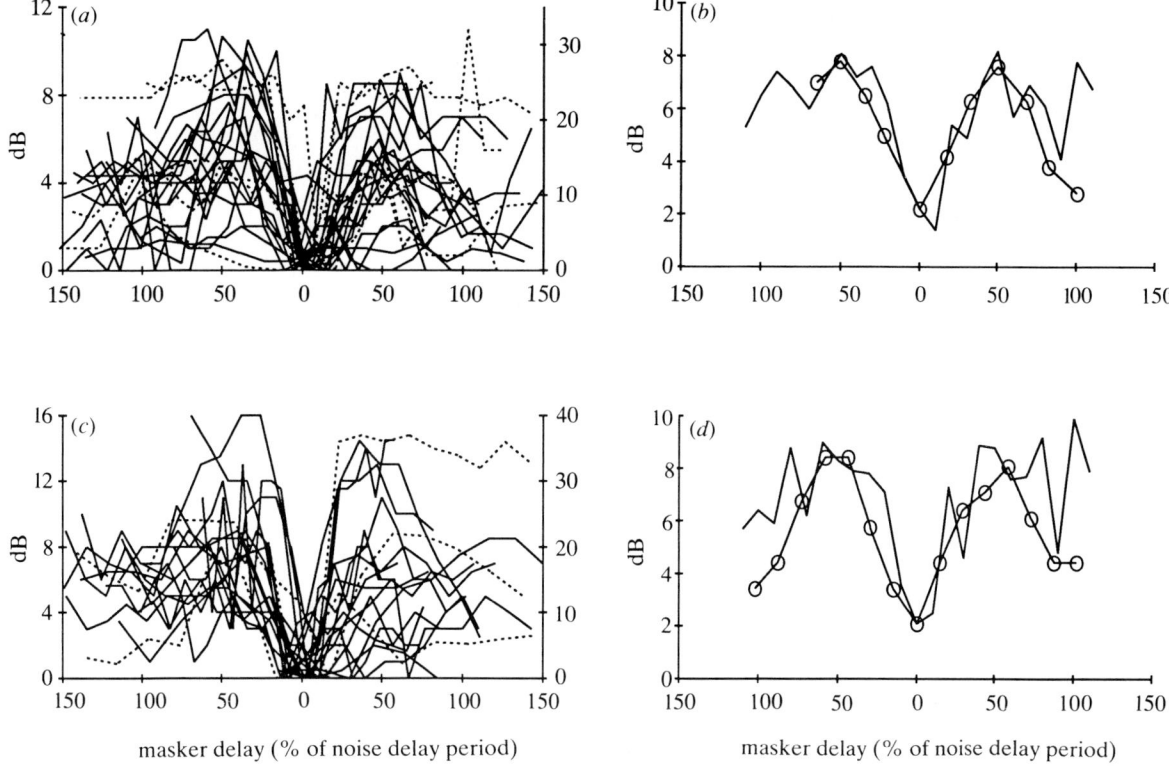

Figure 5. (*a*) The masked thresholds as a function of the masking noise delay for all the cells tested with BF tones (*n* = 25); (*c*) shows similar data for those tested with the vowel stimulus (*n* = 19). The second ordinate refers to the few cells which showed large variations in the masked thresholds (shown as the dotted lines). (*b*) and (*d*) show the average curves obtained from the data in (*a*) and (*c*) respectively computed at intervals of 10% along the abscissa. The open symbols in (*b*) and (*d*) show the psychophysical measures of BMLD (our own vowel data and tone data from Rabiner *et al.* (1966)) under similar conditions which have been scaled and shifted in the same way as the physiological data. (Replotted from figure 6, Caird *et al.* (1991).)

lying the masking, to do this we used a second paradigm. This paradigm is more akin to that which is naturally encountered, in that the level of the noise is constant as its delay is varied; equivalent to the masking of a signal by the same noise at progressively different azimuthal positions. The mean spike rate in the signal plus noise interval, may be a result of excitation by the masking noise and will not necessarily, therefore, be a reliable indicator of the degree to which the noise masker has disrupted the responses to the signal. By using a synthetic vowel we were able to use a single signal irrespective of the best frequency and exploit the ability of collicular neurons to phase-lock to the fundamental of such complex sounds (see Palmer *et al.* 1990) to distinguish the responses to masker and signal.

We have completed the analysis of spike rate and synchronization to a vowel signal on 18 neurons. In nearly all cases, when the masking noise was present, irrespective of its delay, the response to the vowel was reduced, whether assessed in terms of the mean or the synchronized discharge rate. In figure 6 we plot both the mean and the synchronized rates for two cells. For the analyses shown in figure 6*a* the signal (indicated by the arrow) was presented at best delay and the delay of the noise masker varied. For the cell in figure 6*a*, the rate synchronized to the F_0 (triangles) is lowest for noise delay values close to the noise best delay. As

the noise delay is moved away from its best delay, the synchronized rate first increased and then decreases. The synchronized rate function thus approximates a mirror image of the noise delay function (dashes and squares in figure 6). For this cell, therefore, masking, assessed from the synchronized rate, changes in the same direction as the masked thresholds described above.

The data shown in figure 6*b* are from a cell characterized by delay-dependent inhibition. For this cell the signal was located (as shown by the arrow) to evoke activity against which masking could be evaluated. The cell response is so strongly inhibited when the noise is presented at its 'best delay' that both the mean discharge rate and the synchronized rate are lowest at the minimum in the noise delay function. This is consistent with the cell shown in figure 6*a*, but now the most effective masking of the signal does not occur when the noise and signal have the same interaural delay. The change in the cell response is in the opposite sense to the BMLD.

Data from all the cells that we tested with this second paradigm are shown in figure 7. As in figure 5, to emphasize the similarities in the responses, we have eliminated the variation due to differences in the cells' best delay and response frequency. The degree to which the discharge rate was altered by the noise masker varied between neurons (23–100%), on

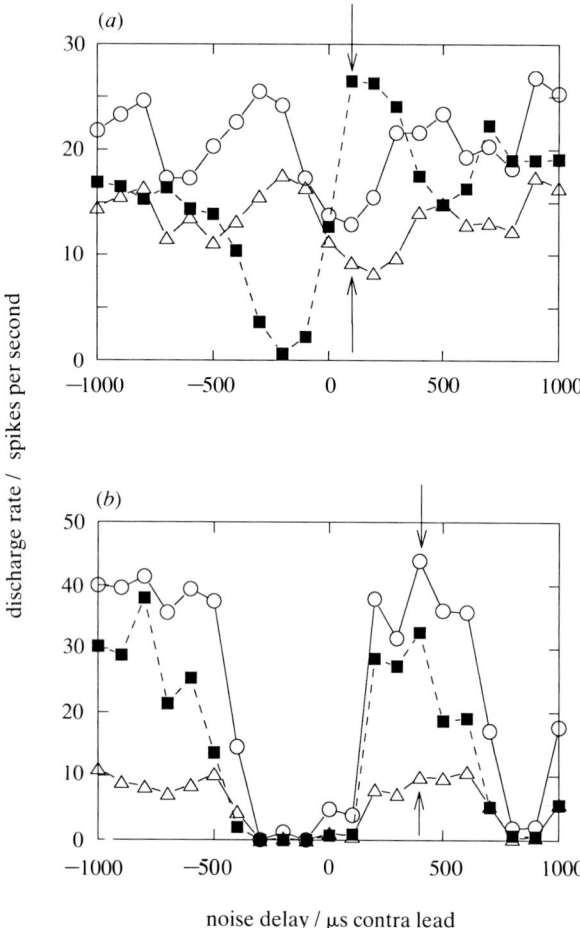

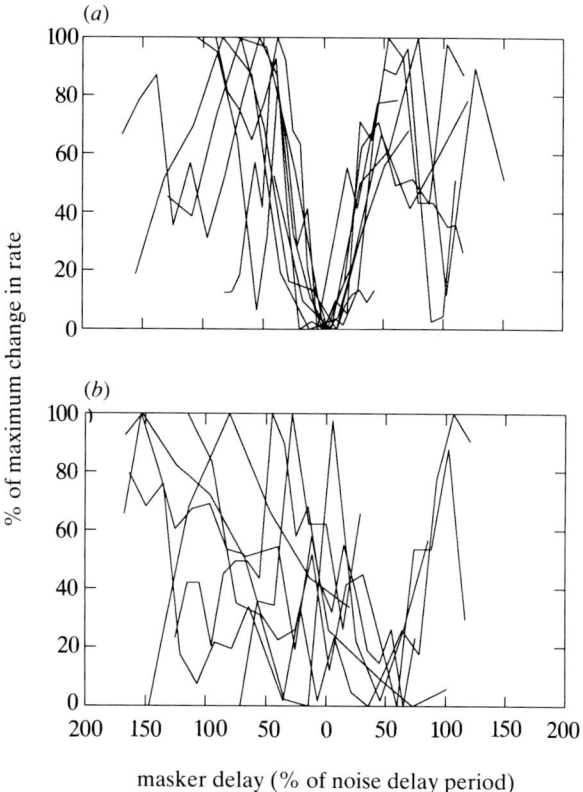

Figure 6. The continuous curves in (*a*) and (*b*) show the mean discharge rate (open circles) and discharge rate synchronized to the fundamental (100 Hz) of the vowel /a/ (open triangles) as a function of the interaural delay of a continuous wideband masking noise. The dashed curve shows the mean discharge to the noise alone. The BFS of the cells were 1500 Hz (*a*) and 990 Hz (*b*). The vowel signal was positioned as indicated by the arrow and at 64 and 54 dB SPL in (*a*) and (*b*) respectively. The level of the noise was adjusted to just completely mask (by audio-visual assessment) the response to the vowel when it was at the best delay (28 dB SPL spectrum level in both cases). (Replotted from figure 7, Caird *et al.* (1991).)

Figure 7. (*a*) Discharge rate synchronized to the F_0 of the vowel versus masker delay for nine cells showing responses like those in figure 6*a*. These data have been normalized for differences in (i) the degree of discharge rate modulation; (ii) the period of the cyclic delay sensitivity, and (iii) the best delay. (*b*) The rate synchronized to F_0 as a function of the masker delay for the remaining seven cells which did not show minima in this response measure at the best delay of the masking noise. (Replotted from figure 9, Caird *et al.* (1991).)

average the response measures changed by 50%. In figure 7, the changes in the synchronized discharge rate as the noise delay is varied are plotted for 16 cells. In figure 7*a* we have grouped together those responses which showed maximum reduction in the synchronized rate when the noise was at its best delay. The responses of the cells in figure 7*a* have maxima near ± 50%. In other words, the synchronous discharge is maximally reduced when the masker is at its best delay and least reduced when the masker is separated by one half cycle of the noise delay functions; an effect that corresponds to the majority of the masked threshold measurements described above. The functions from the rest of our sample are shown in figure 7*b*. The only feature common to these functions is the minimum near the half cycle; the maxima of these curves are widely distributed and appear to bear no obvious relation to the noise delay function.

4. DISCUSSION

(*a*) *Delay sensitivity*

The shapes of the interaural delay functions in the guinea pig, and the positions of their peaks, are very similar to those in the cat (Kuwada & Yin 1983; Yin *et al.* 1986). Indeed, a comparison between our data and those available for other animals shows no significant species differences (rabbit: Aitkin *et al.* 1972; Kuwada *et al.* 1987; kangaroo rat: Moushegian *et al.* 1971; Stillman 1971; barn owl: Moiseff & Konishi 1981). This remarkable similarity of IPD interactions, even across vertebrate classes, strongly suggests that these effects are general rather than species-specific and are a reflection of the physiology of a binaural processing system common to all animals, as has been suggested by other authors (Phillips & Brugge 1985).

(*b*) *Comparison of single unit masking with human* BMLD *measurements*

In any detection task the decision will be made using those neurones which provide the most advantageous signal-to-noise ratio. We have deliberately optimized the BMLD configuration for each cell by using signals placed at the best delay and at a sound level evoking a good response. For any specific BMLD condition there should be a population of cells whose binaural characteristics are optimal for the sound level and interaural delay of the signal. Thus the pooled data in figure 5 may be taken as an indication of the average masking effects which can be detected in the population of cells which have binaural properties matched to the signal.

If we are assessing the responses of a population that is involved in encoding the BMLD, the masking level differences in our study should show a dependence on the stimulus parameters similar to that in the equivalent psychophysical data. Large BMLDs for sinusoids are limited to below 1.5 kHz in humans (see Durlach & Colburn 1978, figure 50). The delay sensitivity of collicular neurons is similarly limited in the guinea pig and other animals, although in the cat behavioural BMLD measures remain large up to 1.5 kHz (Wakeford & Robinson 1974). In human psychophysics, when single sinusoids are masked by wideband noise with different interaural delays, the maximum magnitude of the BMLD is frequency-dependent, with a value of approximately 10 dB at 500 Hz, falling to about 5 dB below 200 Hz and above 1000 Hz (Durlach & Colburn 1978, figure 57). The mean value in our pooled data (figure 5b) is 6.8 dB which falls within the range of values found in humans. Our single cell measures of the masking level difference did not, however, correlate with the frequency of the tone signal. Our own psychophysical data gave a BMLD magnitude for the vowel masked by noise of 7.3 dB. This compares with a value of 6.9 dB in the pooled neuronal data which we obtained using the vowel as the signal (figure 5d).

In psychophysical data the masked threshold cycles as the noise delay is varied. Although such cycling was sometimes evident in the masked thresholds obtained from single neurons (figure 3) it is notably absent from the pooled data obtained with either tone or vowel signals. This is because the masking reflects the noise delay function, which for most neurons is not cyclic (Yin *et al.* 1986; Palmer *et al.* 1990).

Many cells (32%) failed to show delay-dependent masked thresholds, despite being sensitive to the interaural delay of the noise. Carney & Yin (1989) have demonstrated that individual cells, among those sensitive to interaural time delay, fall on a continuum in respect of the strength of the excitatory and inhibitory inputs that they receive from each ear. The cells in the present study that were sensitive to interaural delay and yet showed no BMLD effects, may have levels of excitation and inhibition that are a variable function of delay. The combined effect of these two processes may result in a masked threshold independent of masker delay.

BMLDs can be demonstrated using continuous signals and maskers. Under such masking conditions a comparison between masker alone and masker plus signal is not available within the activity of a single neuron. Our use of a synthetic vowel has allowed the study of information which relates specifically to the signal and

not the noise. The data obtained using this measure show that the degree to which the synchronized response to the vowel is disrupted depends upon the delay sensitivity to the noise. We have shown that both mean rate and temporal patterning of the discharge of a unit are affected by the interaural delay of the masking noise. Although we cannot quantify these changes to allow direct comparisons with psychophysical data, in most of our sample they are in a direction consistent with the BMLD.

We thank Padma Moorjani for assistance throughout this study. Dr J. Culling, Dr I. M. Winter and Professor M. P. Haggard provided very useful comments on the manuscript. The study was supported by the Medical Research Council, the Deutsche Forschungsgemeinschaft (SFB45/B12), by a twinning grant from the European Training Program in Brain and Behavioural Science to A.R.P. and D.C., and a grant from the Small Grants Research Sub-Committee of the University of Newcastle upon Tyne to A.R.

REFERENCES

Aitkin, L.M., Blake, D.W., Fryman, S. & Bock, G.R. 1972 Responses of neurones in the rabbit inferior colliculus. II Influence of binuaral tonal stimulation. *Brain Res.* **47**, 91–101.

Caird, D.M., Palmer, A.R. & Rees, A. 1991 Binaural masking level difference effects in single units of the guinea pig inferior colliculus. *Hear. Res.* **57**, 91–106.

Caird, D.M., Pillman, F. & Klinke, R. 1989 Processing of binaural masking level difference signals in the cat inferior colliculus. *Hear. Res.* **43**, 1–24.

Carney, L.H. & Yin, T.C.T. 1989 Responses of low frequency cells in the inferior colliculus to interaural time differences of clicks: Excitatory and inhibitory components. *J. Neurophysiol.* **62**, 144–161.

Cranford, J.L. 1975 Auditory masking level differences in the cat. *J. comp. Physiol. Psychol.* **89**, 219–233.

Durlach, N.I. & Colburn, H.S. 1978 Binaural phenomena. In *Handbook of perception*, vol. IV (*Hearing*) (eds. E. C. Carterette & M. P. Friedman) pp. 365–466. New York: Academic Press.

Hirsh, I.J. 1948 The influence of interaural phase on interaural summation and inhibition. *J. acoust. Soc. Am.* **20**, 536–544.

Jeffress, L.A. & McFadden, D. 1971 Differences of interaural phase and level in detection and lateralization. *J. acoust. Soc. Am.* **49**, 1169–1179.

Jeffress, L.A., Blodgett, H.C. & Deatherage, B.H. 1962 Masking and interaural phase, II. 167 cycles. *J. acoust. Soc. Am.* **28**, 1124–1126.

Kuwada, S., Stanford, T.R. & Batra, R. 1987 Interaural phase-sensitive units in the inferior colliculus of the unanaesthetized rabbit: effects of changing frequency. *J. Neurophysiol.* **57**, 1338–1360.

Kuwada, S. & Yin, T.C.T. 1983 Binaural interaction in low-frequency neurons in inferior colliculus of the cat. I. Effects of long interaural delays, intensity, and repetition rate on interaural delay function. *J. Neurophysiol.* **50**, 981–999.

Langford, T.L. 1984 Responses elicited from medial superior olivary neurons by stimuli associated with binaural masking and unmasking. *Hear. Res.* **15**, 39–50.

Licklider, J.C.R. 1948 The influence of interaural phase relations upon the masking of speech by white noise. *J. acoust. Soc. Am.* **20**, 150–159.

Moiseff, A. & Konishi, M. 1981 Neuronal and behavioural sensitivity to binaural time differences in the owl. *J. Neurosci.* **1**, 40–48.

Moushegian, G., Stillman, R.D. & Rupert, A.L. 1971 Characteristic delays in superior olivary complex and inferior colliculus. In *Physiology of the auditory system* (ed. M. B. Sachs), pp. 245–254. Baltimore: National Educational Consultants.

Palmer, A.R., Rees, A. & Caird, D.M. 1990 Interaural delay sensitivity to tones and broad band signals in the guinea-pig inferior colliculus. *Hear. Res.* **50**, 71–86.

Palmer, A.R., Winter, I.M. & Darwin, C.J. 1986 The representation of steady-state vowel sounds in the temporal discharge patterns in the guinea pig cochlear nerve and primary like cochlear nucleus neurons. *J. acoust. Soc. Am.* **79**, 100–113.

Phillips, D.P. & Brugge, J.F. 1985 Progress in neurophysiology of sound localization. *A. Rev. Psychol.* **36**, 245–274.

Rabiner, L.R., Laurence, C.L. & Durlach, N.I. 1966 Further results on binural unmasking and the EC model. *J. acoust. Soc. Am.* **40**, 62–70.

Stillman, R.D. 1971 Characteristic delay neurons in the inferior colliculus of the kangaroo rat. *Expl Neurol.* **32**, 404–412.

Taylor, M.M. & Creelman, C.D. 1967 PEST: Efficient estimates of probability functions. *J. acoust. Soc. Am.* **41**, 782–787.

Wakeford, O.S. & Robinson, D.E. 1974 Detection of binaurally masked tones by the cat. *J. acoust. Soc. Am.* **56**, 952–956.

Yin, T.C.T., Chan, J.C.K. & Carney, L.H. 1987 Effects of interaural time delays of noise stimuli on low-frequency cells in the cat's inferior colliculus. III. Evidence for cross-correlation. *J. Neurophysiol.* **58**, 562–583.

Yin, T.C.T., Chan, J.C.K. & Irvine, D.R.F. 1986 Effects of interaural time delays of noise stimuli on low-frequency cells in the cat's inferior colliculus. I. Responses to wide-band noise. *J. Neurophysiol.* **55**, 280–300.

Philosophy and stimulus design for neuroethology of complex-sound processing

NOBUO SUGA

Department of Biology, Washington University, Lindell-Skinker Avenue, St Louis, Missouri 63130, U.S.A.

SUMMARY

In research on the neural mechanisms for the processing of biologically important sounds such as species-specific sounds and sounds produced by prey and predators, it is necessary to study responses of central auditory neurons to biologically important sounds, information-bearing elements (IBES) in them, and tone bursts. The tone bursts or constant-frequency (CF) components can be an IBE in many species of animals. Information-bearing parameters characterizing these sounds must be systematically varied, and tuning of neurons to individual parameters must be studied. The measurement of a tuning curve must be performed not only for excitatory responses, but also for inhibitory and facilitative responses, if any. The selectivity of a neuron to a particular type of sound must be tested for whether it is level-tolerant. Responses to complex sounds can probably be explained on the basis of those to IBES and tone bursts, so that the use of the tone bursts, even though they are not IBES, is as essential as that of the biologically important sounds.

1. INTRODUCTION

The processing of communication calls is one of the major functions of the central auditory system. However, nothing interesting has been found of the central processing of calls in mammalian species, in spite of the promising preliminary research made in the early 1970s (Wollberg & Newman 1972), and elegant measurements on frequency tuning, amplitude tuning, binaural interactions, phase-locking, etc. over the past 25 years. On the other hand, noticeable progress has been made in birds (Margoliash 1983; Müller & Leppelsack 1985; Scheich *et al.* 1979) and frogs (Mudry *et al.* 1977; Fuzessery & Feng 1983). In songbirds and frogs, it has been shown that species-specific calls are processed by neurons that are tuned to particular combinations of two signal elements. This is also true for the processing of biosonar information in bats (see Suga *et al.* 1983). Although many more analytical and comparative studies are needed, it is clear that all of these three types of animals (bats, birds and frogs) share the same principle: species-specific complex sounds, either communication calls or biosonar signals, are processed by combination-sensitive neurons. In bats (see Suga *et al.* 1983) and frogs (Hall & Feng 1987), the central auditory system creates separate clusters of neurons tuned to different types of information-bearing parameters. These findings indicate that a neuroethological approach has been successful in the exploration of neural mechanisms for the processing of species-specific complex sounds. As I have written several articles reviewing our research on the mustached bat (for examples see Suga 1984, 1988, 1990), I shall review here the philosophy and stimulus design for auditory neuroethology.

2. INFORMATION-BEARING ELEMENTS AND PARAMETERS

In higher vertebrates, communication sounds are usually complex, and the amplitude spectra of these sounds commonly change with time. For example, human speech consists of various phonemes combined in different sequences and is consequently complicated. However, the spectrograms of speech sounds exhibit three basic components (information-bearing elements: IBES): constant frequency (CF), noise burst (NB), and frequency-modulated (FM) components. For example, consider the consonant-vowel syllable of figure 1. At the beginning of the syllable, there is a

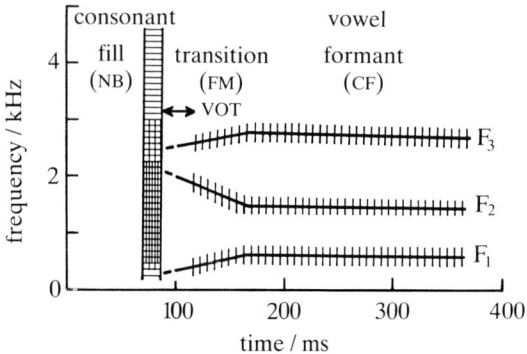

Figure 1. Information-bearing elements in human speech sounds. Schematized spectrogram of a monosyllabic sound shows four types of information-bearing elements: fill or noise burst (NB) component, transition or frequency modulated (FM) component, formant or constant frequency (CF) component, and voice onset time (VOT) or time interval between two acoustic events. F_1, F_2 and F_3 are the first, second and third formants of a vowel, respectively. The short vertical bars on F_1, F_2 and F_3 indicate vocal chord activity (Suga 1984; based on Liberman *et al.* 1956).

Phil. Trans. R. Soc. Lond. B (1992) **336**, 423–428
Printed in Great Britain

423

© 1992 The Royal Society

vertical bar (NB component) as the air is released from the vocal tract; this corresponds to the NB portion of the consonant. The last 200 ms of the sound (the vowel portion) consists of several horizontal bars, called formants or CF components. Between the burst and vowel, there are oblique bars, called transitions or FM components (cf. Liberman *et al.* 1956).

Auditory information is carried not only by the acoustic parameters characterizing each of the above three types of IBES, but also by parameters representing relationships among these IBES in the frequency, amplitude and time domains. Values of a parameter comprise a continuum, but only a limited portion of the continuum is important for each species. This portion has been called the information-bearing parameter: IBP (Suga *et al.* 1981).

The three types of IBES are also found in sounds produced by many different types of animals (Suga 1972). For example, the mustached bat, *Pteronotus*

parnellii, emits complex biosonar pulses consisting of CF and FM components (figure 2*a*). These are IBES carrying different types of biosonar information. It also emits a variety of communication sounds that differ from biosonar pulses in amplitude spectrum and duration (e.g. figure 2*b–j*). The central auditory system of different species of animals contains not only neurons selectively responding to one of the IBES (see Suga 1969, 1973), but also neurons tuned to particular combinations of IBES (in bats, see Suga *et al.* 1978, 1983; in birds, Margoliash 1983; in frogs, Fuzessery & Feng, 1983). In the mustached bat, arrays of combination-sensitive neurons tuned to different values of IBPS are systematically arranged in the auditory cortex and thus form computational maps (e.g. O'Neill & Suga 1982; Suga & O'Neill 1979; Suga *et al.* 1983). In the barn owl, *Tyto alba*, neurons tuned to different values of interaural time and amplitude differences are systematically arranged in the external nucleus of the inferior colliculus and thus

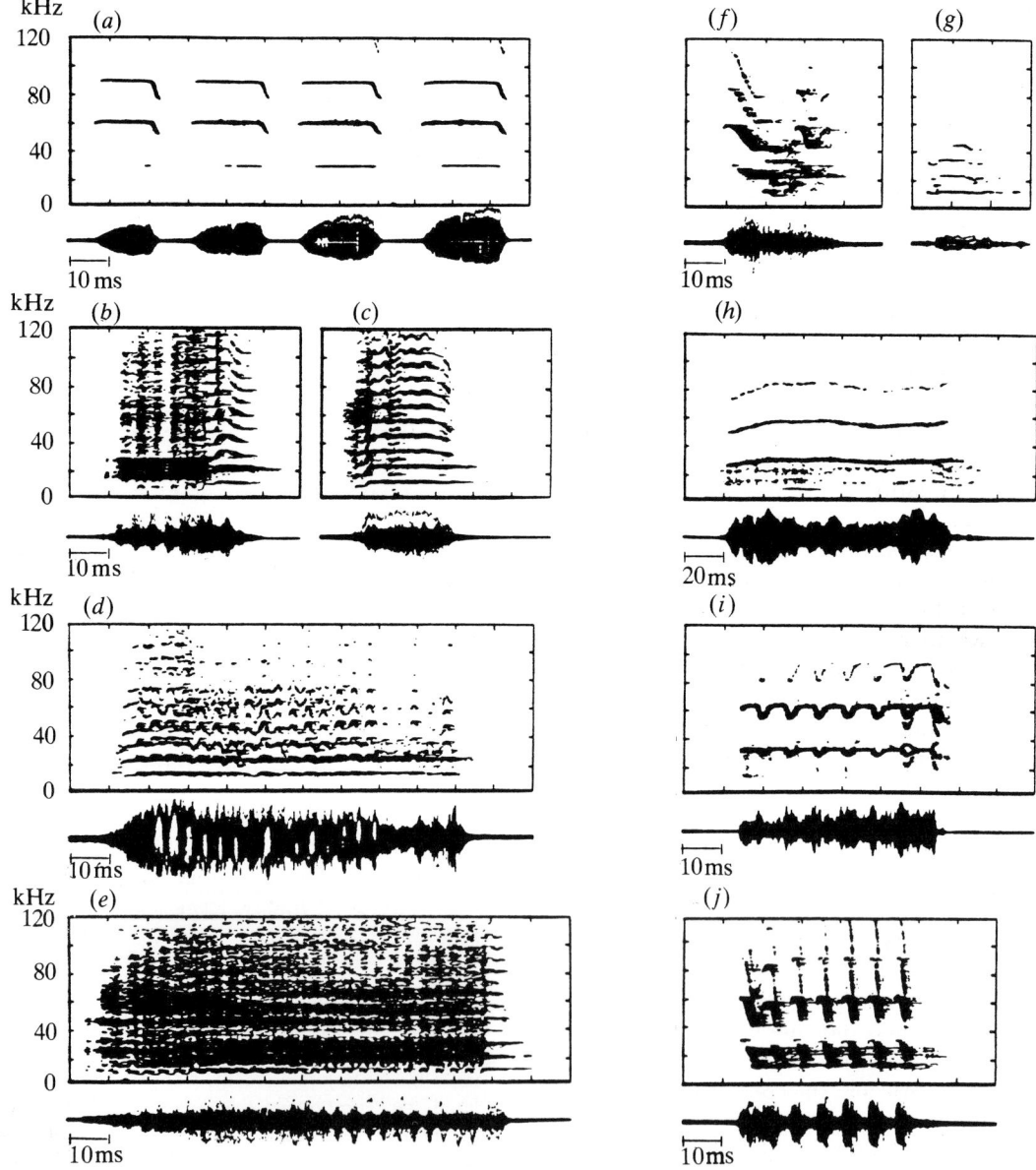

Figure 2. Biosonar pulses (*a*) and communication calls (*b–j*) of mustached bats. Note the clear differences in amplitude spectrum and temporal pattern between the pulses and the calls, except call (*j*).

form an auditory space map (Knudsen & Konishi 1978). The data obtained from the mustached bat and the barn owl clearly indicate that the central auditory system creates neurons tuned to different values of an IBP and form an IBP map. These interesting data were obtained from neuroethological studies of the auditory system.

3. PHILOSOPHY BEHIND AUDITORY NEUROETHOLOGY

Which do we want to explore, neural mechanisms for the processing of species-specific sounds or those for the processing of 'general' sounds? The auditory system has evolved for processing biologically important sounds, in particular, species-specific sounds and sounds produced by prey and predators. Therefore, it is reasonable to explore first the mechanisms for processing biologically important sounds, especially, species-specific sounds rather than those for the processing of general sounds.

In general, animals appear to have subsets of neurons specialized for processing species-specific sounds. Finding such neurons and exploring the neural mechanisms for creating them require the delivery of the species-specific sounds, individual IBEs found in them, and tone bursts. (Tone bursts can be an IBE in many species of animals.) When a neuron responds selectively to a particular complex sound, we must first confirm that no single component in the sound, whatever its amplitude, is sufficient to excite significantly the neuron, and then examine (i) which combination of two or more components is necessary for the excitation of the neuron, and (ii) how sharply the neuron is tuned to particular IBPs characterizing single IBEs or combinations of IBEs. The measurement of the filter properties of the neuron is particularly important for a quantitative description of its response properties, and also for a correlation of neural responses with biological sounds that always show some variation. Since communication, as well as biosonar, can occur over different distances, it is also important to examine whether the selectivity of the neuron to a particular sound is 'level-tolerant', i.e. does not change with stimulus level. It is particularly important to explore the neural mechanisms creating the selectivity, i.e. interaction among excitation, inhibition and facilitation.

4. STIMULUS DESIGN TO CHARACTERIZE NEURONS

In the past, species-specific calls were recorded with a tape-recorder or a computer and were played back with modifications to assess important portions of the calls to excite neurons. The modifications were made by either trimming the calls from their beginning or end, or playing them backward. Using such acoustic stimuli, some neurons were found to respond selectively to a particular 'natural' call, but not to either the reversed call or tone bursts (CF tones). Such research, however, is not analytical enough to characterize neurons. If a neuron responds only to a natural call or

a particular portion of the call, its selectivity must be tested for level-tolerance; the call element or elements which excite the neuron should be identified; how sharply the neuron is tuned to a parameter or parameters characterizing the call element or elements should be measured; and the reasons why the neuron responds to call elements, but not to tone bursts must be determined. For research on the neural processing of species-specific complex calls, the calls are first categorized, and 'typical' calls are used as acoustic stimuli representing the call types. Within each call type, however, individual calls may show large variation in the frequency, amplitude and time domains. Therefore, the typical call stimuli must be varied in these three domains. To fulfill all of the above requirements, acoustic stimuli must be carefully designed.

There are two different approaches to the design of acoustic stimuli. These differ from each other with respect to which set of acoustic stimuli is to be first used, IBEs or natural calls. In one approach, IBEs and combinations of IBEs found in species-specific calls are synthesized with special-purpose instruments or a computer (figure 3), and their IBPs are systematically varied, whereas responses of single neurons are recorded. If IBEs in a call are unknown, these can be speculated by examining the spectrogram of the call. Based upon what is known about speech processing, these are usually predominant components. The IBEs speculated can then be synthesized, and their parameters can be systematically varied. Natural calls are eventually used as stimuli to examine responses of neurons.

In the other approach, typical calls are used as acoustic stimuli with or without modifications by a computer. Modifications include amplification or attenuation, filtering, shifting in frequency, cutting,

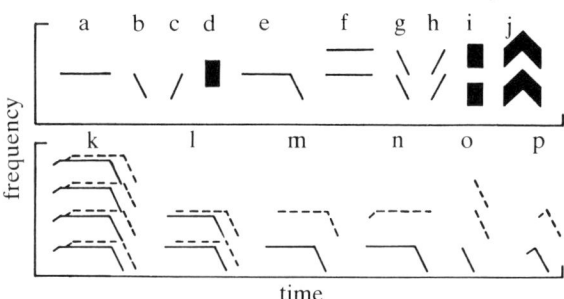

Figure 3. Schematized spectrograms of the acoustic stimuli which were used in our experiments on the biosonar of the mustached bat and the little brown bat. (*a*) CF tone (tone burst); (*b*) downward-sweeping FM sound; (*c*) upward-sweeping FM sound; (*d*) noise burst; (*e*) CF–FM sound; (*f, g, h* & *i*) correspond respectively to signals in (*a, b, c* & *d*), but with overtones; (*j*) noise burst in which the centre frequency first increases and then decreases; (*k*) a pair of a biosonar pulse (solid lines) and its Doppler-shifted echo (dashed lines); (*l–p*) five examples of simplified biosonar pulse (solid line) and its echo (dashed line). The delay of the echo can be varied. The duration, intensity and frequency of each signal (or component) can be varied. In FM signals, frequency can be swept linearly, exponentially, sinusoidally or trapezoidally with time. The range of the frequency sweep can also be changed.

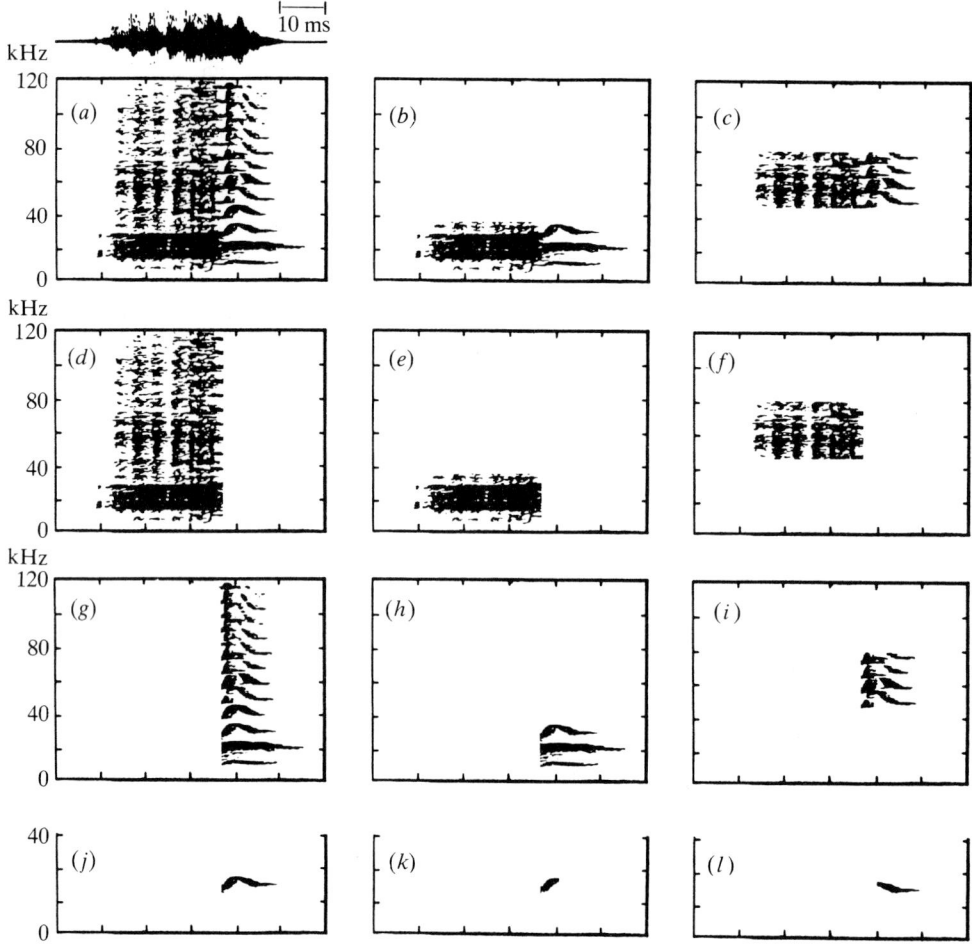

Figure 4. Modifications of a communication call to examine which part of the call is essential to excite a 'specialized' neuron. The trace at the top left shows the envelope of the call in (*a*). The normal call (*a*) is modified in the frequency domain (*b* or *c*) or time domain (*d* & *g*) or frequency-time domains (*e, f, h* & *i*). If the component shown in (*j*) excites the neuron, it will be further tested whether the neuron is sensitive to an upward-sweep (*k*) or downward-sweep (*l*) or their combination (*j*). If the neuron is sensitive to the combination, the tuning curve of the neuron will be measured as a function of each parameter characterizing the upward-sweep, downward-sweep, or the combination of these.

pasting, etc. For studies of single neurons, quick modifications of acoustic stimuli are required, so that filtering of the calls in the frequency and time domains can, respectively, be performed with a programmable filter which has a steep slope (e.g. 115 dB per octave) and an electronic switch which can independently control the delay, duration and rise-decay time (figure 4). Alternatively, a large file of modified acoustic stimuli can be stored in a computer. After identifying the essential signal elements for excitation of a neuron, the elements can be synthesized, and their parameters can be systematically varied.

The first approach has been taken in the studies on the auditory system of the mustached bat (e.g. Olsen & Suga 1991*a, b*; O'Neill & Suga 1982; Suga & Tsuzuki 1985; Suga *et al.* 1978, 1983), white-crowned sparrow (Margoliash 1983), and frog (Fuzessery & Feng 1982, 1983). All these studies have demonstrated that species-specific sounds or sequenced sounds are processed by 'combination-sensitive' neurons, and that they are clustered within certain portions of the brain.

5. EXPLORATION OF NEURAL MECHANISMS FOR PROCESSING COMPLEX SOUNDS

In the past, experiments on the auditory system of avians and mammals, communication calls of a species were recorded with a tape recorder or a computer and were played back to members of that species at, say, 60 dB SPL. Responses of single auditory neurons to the calls were compared with those to tone bursts (CF tones) at 60 dB SPL. Then, some neurons were found to be excited only by a particular type of call, but to be inhibited by the tone bursts. Therefore, it was concluded that these neurons were specialized to respond to the call, and that the responses to the call could not be explained by the responses to the tone bursts. These conclusions, however, are unacceptable because of the following two major reasons: (i) the 60 dB call (broad band signal) and 60 dB tone burst (narrow band signal) are the same in intensity, but the tone burst is much stronger than the call component alone corresponding to the tone burst (figure 5*a*); and (ii) the response properties of many central

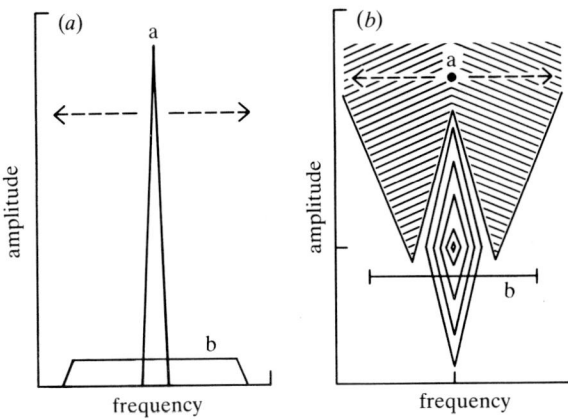

Figure 5. Amplitude spectra of two acoustic stimuli (*a*) and a response area of a central auditory neuron (*b*). (*a*) A tone burst (a) and a broadband signal (b) are the same in intensity. The frequency of the tone burst can be varied as indicated by the dashed arrows. (*b*) The excitatory and inhibitory areas of the neuron are indicated by the largest diamond and a shaded area, respectively. The four diamonds within the excitatory area indicate the iso-impulse-count contour lines. The neuron is tuned in frequency and amplitude (based upon Suga & Manabe (1982)). The neuron shows only an inhibitory response to the tone burst shown in (*a*), regardless of its frequency, because it stimulates only the inhibitory area. However, it shows excitatory response to the broadband signal shown in (*a*), because it stimulates only the excitatory area.

auditory neurons are extremely nonlinear (figure 5*b*). Therefore, it is incorrect to compare the response to the 60 dB call with that to the 60 dB tone burst. For example, some central neurons are tuned to a combination of a particular frequency and a particular amplitude and show 'upper-threshold' (in frogs, Potter 1965; in bats, Grinnell 1963; Suga 1977; Suga & Manabe 1982; in cats, Phillips & Orman 1984). 'Upper-threshold' neurons show inhibitory responses to strong tone bursts, but excitatory responses to weak tone bursts (figure 5*b*; in frogs, Fuzessery & Feng 1982; in bats, Suga 1965; Suga & Manabe 1982). Then, the response properties of these neurons characterized with tone bursts can be used to explain the observation that the 60 dB tone burst (a in figure 5*b*) does not excite but inhibits the neuron, whereas the 60 dB call (b in figure 5*b*) excites the neuron. The characterization of response properties of central auditory neurons only at a particular stimulus level is misleading, so that their response properties must be examined at different stimulus levels.

There are other examples which do not favour the statement that the responses to complex calls cannot be explained on the basis of responses to tone bursts. FM specialized neurons are inhibited by tone bursts, but excited by FM sounds which sweep across the inhibitory area. The response properties of these neurons can be explained by a disinhibition model, based upon responses to tone bursts (Suga 1965, 1968, 1969). Some cortical CF/CF and FM–FM combination-sensitive neurons do not show excitatory response to any single tone burst and FM sound, but show excitatory responses to particular combinations of a

pulse and an echo (see, for example, Suga *et al.* 1983). The response properties of these CF/CF and FM–FM neurons can be explained by facilitation of responses taking place in the medial geniculate body, in other words, by responses to tone bursts (Olsen & Suga 1991*a, b*).

All peripheral neurons, without exception, respond to CF tones, FM sounds and noise bursts. Response properties of central auditory neurons are based upon excitatory, inhibitory and facilitatory interactions in the frequency, amplitude and time domains among neurons receiving signals directly or indirectly from these peripheral neurons. Therefore, the neural mechanisms for creating the responses of neurons selective to particular types of calls can be explored by step-by-step analysis of synaptic interactions occurring at different levels of the auditory system with the three types of IBES (CF tones, FM sounds and noise bursts) which are commonly found in animal sounds.

To explore the neural mechanisms for the creation of 'specialized' neurons that respond selectively or preferentially to particular types of complex sounds, their response properties, in particular, tuning curves, must be extensively studied. Because responses of these neurons probably consist of excitation, inhibition and facilitation which vary in frequency, amplitude and time, the properties of these synaptic events must be studied in these three domains. Excitatory responses and tuning curves are studied with single IBES (e.g. tone bursts). Inhibitory responses and tuning curves are studied with single, double or triple IBES (e.g. Suga 1965; Suga & Tsuzuki 1985). When the background activity of a neuron is inhibited by single sounds, inhibition is studied with the single sounds. When it is absent or is not inhibited by single sounds, inhibition is studied with paired sounds. One sound is delivered before, without overlap with, a second sound which excites the neuron. An inhibitory tuning curve is measured as the frequency and amplitude ranges of the first sound that causes inhibition of the response to the second. The timecourse of the inhibition is measured by changing the time relationship between the two sounds. In combination-sensitive or call-sensitive neurons, the second sound in the pair is a 'combination' sound or a call that excites the neuron. Facilitative responses or tuning curves are studied with a pair of sounds (e.g. Suga *et al.* 1978, 1983). One sound in the pair is fixed in frequency and amplitude for optimum facilitation, while the other is varied in these parameters to measure a facilitative tuning curve, or one sound is delayed from the other to study the time-course of facilitation. When a neuron is multi-combination-sensitive or call-sensitive, one of the paired sounds is a fixed complex sound or a simplified call essential for evoking facilitation, while the other is varied in frequency, amplitude and time.

Complete studies of response properties of single neurons require stable recording of action potentials at least over a few hours. Since stable intracellular recording over the long time period is hard, extracellular recording is preferred to intracellular recording. The characterization of single neurons in extracellular recording tells us the end result of neural interactions taking place at all levels interposed between peripheral neurons and a given central neuron. It does not tell us where and how particular neural interactions take place. Therefore, a top-down study must be performed step-by-step. Differences in response properties between cortical areas or subcortical nuclei and the knowledge of anatomical connections give us important insights into the neural mechanisms for processing complex sounds within and across cortical areas and sub-cortical nuclei.

For further exploration of the neural mechanisms for processing complex sounds or creating specialized neurons, synaptic mechanisms occurring in each step must be explored by micro-iontophoretic injections of various drugs with multi-barrelled electrodes. In such studies, response properties of a single neuron are first studied as described above, and drugs such as synaptic transmitters, their agonists, and their antagonists are injected. Response properties of the neuron are re-studied during and after injections. The changes in response properties evoked by drugs are evaluated in relation to working hypotheses.

Our research on the bat's auditory system has been supported by the research grants from NIDCD (DC00175), ONR (N00014-90-J-1068) and the McKnight Foundation. I thank J. S. Kanwal and K. K. Ohlemiller for their comments on the present article.

REFERENCES

Fuzessery, Z.M. & Feng, A.S. 1982 Frequency selectivity in the anuran auditory midbrain: single unit responses to single and multiple tone stimulation. *J. comp. Physiol.* A **146**, 471–484.

Fuzessery, Z.M. & Feng, A.S. 1983 Mating call selectivity in the thalamus and midbrain of the leopard frog *Rana p. pipiens*: single and multiunit analysis. *J. comp. Physiol.* **150**, 333–344.

Grinnell, A.D. 1963 The neurophysiology of audition in bat: intensity and frequency parameters. *J. Physiol., Lond.* **167**, 38–66.

Hall, J.C. & Feng, A.S. 1987 Evidence for parallel processing in the frog's auditory thalamus. *J. comp. Neurol.* A **258**, 407–419.

Knudsen, E.I. & Konishi, M. 1978 A neural map of auditory space in the owl. *Science, Wash.* **200**, 795–797.

Liberman, A.M., Delattre, P.C., Gerstman, L.J. & Cooper, F.S. 1956 Tempo of frequency change as a cue for distinguishing classes of speech sounds. *J. exp. Psychol.* **52**, 127–137.

Margoliash, D. 1983 Acoustic parameters underlying the responses of song-specific neurons in the white-crowned sparrow. *J. Neurosci.* **3**, 1039–1057.

Mudry, K.M., Constantin-Paton, M. & Caprinica, R.R. 1977 Auditory sensitivity of the diencephalon of the leopard frog *Rana p. pipiens*. *J. comp. Physiol.* **114**, 1–13.

Müller, C.M. & Leppelsack, H.J. 1985 Feature extraction and tonotopic organization in the avian auditory forebrain. *Expl Brain Res.* **59**, 589–599.

Olsen, J.F. & Suga, N. 1991*a* Combination-sensitive neurons in the medial geniculate body of the mustached bat: encoding of relative velocity information. *J. Neurophysiol.* **65**, 1254–1274.

Olsen, J.F. & Suga, N. 1991*b* Combination-sensitive neurons in the medial geniculate body of the mustached bat: Encoding of target range information. *J. Neurophysiol.* **65**, 1275–1296.

O'Neill, W.E. & Suga, N. 1982 Encoding of target-range information and its representation in the auditory cortex of the mustached bat. *J. Neurosci.* **47**, 225–255.

Phillips, D.P. & Orman, S.S. 1984 Responses of single neurons in posterior field of cat auditory cortex to tonal stimuli. *J. Neurophysiol.* **51**, 147–163.

Potter, D. 1965 Patterns of acoustically evoked discharges of neurons in the mesencephalon of the bullfrog. *J. Neurophysiol.* **28**, 1155–1184.

Scheich, H., Langner, G. & Bonke, D. 1979 Responsiveness of units in the auditory neostriatum of the guinea fowl *Humida meleagris* to species-specific calls and synthetic stimuli. II: Discrimination of iambus-like calls. *J. comp. Physiol.* **132**, 257–276.

Suga, N. 1965 Functional properties of auditory neurons in the cortex of echolocating bats. *J. Physiol., Lond.* **181**, 671–700.

Suga, N. 1968 Analysis of frequency-modulated and complex sounds by single auditory neurons of bats. *J. Physiol., Lond.* **198**, 51–80.

Suga, N. 1969 Classification of inferior collicular neurons of bats in terms of responses to pure tones, FM sounds and noise bursts. *J. Physiol., Lond.* **200**, 555–574.

Suga, N. 1972 Analysis of information-bearing elements in complex sounds by auditory neurons of bats. *Audiology* **11**, 58–72.

Suga, N. 1973 Feature extraction in the auditory system of bats. In *Basic mechanisms in hearing.* (ed. A. R. Möller), pp. 675–744. New York: Academic Press.

Suga, N. 1977 Amplitude-spectrum representation in the Doppler-shifted-CF processing area of the auditory cortex of the mustache bat. *Science, Wash.* **196**, 64–67.

Suga, N. 1984 The extent to which biosonar information is represented in the bat auditory cortex. In *Dynamic aspects of neocortical function.* (ed. G. M. Edelman, W. E. Gall & W. M. Cowan), pp. 315–373. New York: John Wiley & Sons.

Suga, N. 1988 Auditory neuroethology and speech processing: complex sound processing by combination-sensitive neurons. In *Functions of the auditory system* (ed. G. M. Edelman, W. E. Gall & W. M. Cowan), pp. 679–720. New York: John Wiley & Sons.

Suga, N. 1990 Cortical computational maps for auditory imaging. *Neural Networks* **3**, 3–21.

Suga, N., Kujirai, K. & O'Neill, W.E. 1981 How biosonar information is represented in the bat cerebral cortex. In *Neuronal mechanisms of hearing* (ed. J. Syka & L. Aitkin), pp. 197–219. New York: Plenum.

Suga, N. & Manabe, T. 1982 Neural basis of amplitude-spectrum representation in the auditory cortex of the mustached bat. *J. Neurophysiol.* **47**, 225–255.

Suga, N. & O'Neill, W.E. 1979 Neural axis representing target range in the auditory cortex of the mustached bat. *Science Wash.* **206**, 351–353.

Suga, N., O'Neill, W.E., Kujirai, K. & Manabe, T. 1983 Specificity of combination-sensitive neurons for processing of complex biosonar signals in the auditory cortex of the mustached bat. *J. Neurophysiol.* **49**, 1573–1626.

Suga, N., O'Neill, W.E. & Manabe, T. 1978 Cortical neurons sensitive to particular combination of information-bearing elements of biosonar signals in the mustached bat. *Science, Wash.* **200**, 778–781.

Suga, N. & Tsuzuki, K. 1985 Inhibition and level-tolerant frequency tuning in the auditory cortex of the mustached bat. *J. Neurophysiol.* **53**, 1109–1145.

Wollberg, Z. & Newman, J.D. 1972 Auditory cortex of the squirrel monkey: response patterns of single cells to species-specific vocalizations. *Science Wash.* **175**, 212–214.

Index